Don Fifie

Water Pollution

Environmental Issues Series
Scientists' Institute for Public Information

Edited by Barry Commoner

Already published in the series:

Air Pollution

Water Pollution

Julian McCaull

Editor,
Environment

Janice Crossland

Research Associate
Environment

Harcourt Brace Jovanovich, Inc.
New York Chicago San Francisco Atlanta

Cover: "Polluted Lake Zug in Switzerland." Photograph by Georg Gerster, Rapho-Guillumette.
Photo credits and copyright acknowledgments appear on page 199.

ISBN: 0–15–595125–4

Library of Congress Catalog Card Number: 73–13334

Printed in the United States of America

Preface

Water pollution is obvious enough to all of us. The immediate cause seems apparent—no farther away than the discharge pipes of the nearest industrial or sewage treatment plant. The underlying reasons for the continued use of these pipes have resisted definition, however. In this book we explain that the use of discharge pipes, first a matter of convenience, eventually generated a technological rationale that has now achieved formal status. Our belief is that lasting solutions to water pollution require a reassessment of this rationale, particularly because many current conventional water pollution control and water use concepts are derived from nineteenth-century practices. *Water Pollution* is designed to help in this reassessment process, which should occur at the college, graduate, and professional levels of education where the rationale receives attention.

The scientific subjects discussed do not require students to have any previous background. The social problems associated with the many facets of water pollution are related to scientific and technological aspects, thus enabling students to identify the links between the social and the scientific issues inherent in pollution—issues that must be resolved if the situation is to be improved. Many of the issues closely parallel those raised by Virginia Brodine in *Air Pollution,* a companion book in this series.

We are grateful to the many persons who have helped us. This book would not have been possible without the continuous assistance of Virginia Brodine, who first helped us organize and develop the manuscript and then oversaw the organizational details to be coordinated in such a project. Virginia did a good deal of rewriting and much of the final editing; she also took on the intricate task of selecting illustrations. Those of us who worked on *Water Pollution* express our deep appreciation for her efforts.

The authors assume all responsibility for the facts and conclusions presented in the book, and express appreciation for the advice and criticism of the professional people who reviewed the manuscript or sections of it: Congressman John A. Blatnik, Chairman, Public Works Committee, U. S. House of Representatives; John Cairns, Jr., Director and University Professor of Zoology, Center for Environmental Studies, Virginia Polytechnic Institute; J. R. Christensen, Professor of Microbiology, The University of Rochester School of Medicine and Dentistry; Sherburne F. Cook, Jr., Science Program Coordinator, Evergreen State College, Washington; John Dobbs, John Dobbs & Associates, consultants in waste utilization; W. T. Edmondson, Professor of Zoology, University of Washington; M. Grant Gross, National Science Foundation; G. Fred Lee, Professor of Civil Engineering and Head, Environmental Chemistry Program, Environmental Engineering-Environmental Science Division, Texas A&M University; Jack Neff, Associate Professor of Molecular Biology, Vanderbilt University; John Noell, Research Associate, Botany Department, University of Illinois; Thomas Page, Aquatic Physiologist, Ecosystems Department, Battelle Northwest; David J. Wilson, Professor of Chemistry, Vanderbilt University; and Howard Zar, Physical Scientist, U.S. Environmental Protection Agency, Region V.

Many conceptual ideas and much valuable criticism were contributed by Kevin Shea and Sheldon Novick, colleagues at *Environment* magazine. Dianne Hoener, Tina Rehg, and Patricia DeJoie assisted Virginia Brodine in picture research. The demanding jobs of typing and retyping the manuscript and of organizing the review procedures were skillfully performed by Diane Sullivan.

Julian McCaull
Janice Crossland

Foreword

The problems of environmental degradation—air and water pollution, solid waste, threats to wildlife, destruction of natural resources—are often summed up in the phrase "the environmental crisis." One of the chief justifications for using such a phrase is the view that all these problems add up to a qualitative change in the status of the environment, a break with the past serious enough to merit the term "crisis." Against this view some people argue that there is no real crisis, that environmental pollution today—although bad and undesirable—is nothing new, and that in some ways the situation is better than it was.

This argument is worth a closer look, and the problems of water pollution afford an excellent opportunity. It is useful to compare the water pollution situation in the United States today with the situation, let us say, 100 years ago.

Of all the varied water pollution problems discussed in this book—bacterial contamination, the dumping of organic wastes that deplete oxygen content, algal overgrowth, industrial chemicals, thermal pollution—only one was important in the nineteenth century. In the 1870s the single, overwhelming cause of water pollution was the contamination of surface waters, especially drinking water, with the pathogenic bacteria originating in human waste. The most serious bacterial contamination caused typhoid fever.

Certainly on its face the typhoid problem of the 1870s and 1880s seems much worse than any present water pollution problem. As pointed out in Chapter 6, at that time typhoid was responsible for a death rate of about 75–100 per 100,000 population, and most of the deaths were due to contaminated water supplies. This problem has certainly been solved (the death rate from typhoid is now about 0.1 per 100,000), and surely no other present water pollution problem appears to cause as much harm to human health as typhoid did 100 years ago.

This seems to support the argument that water pollution is an old problem, once a good deal worse than it is at present, and that there is no reason to speak of an "environmental crisis" now. On closer examination, however, matters do not appear to be so simple.

Pollution of water by typhoid bacteria created the hazard of a very specific disease; it had a single cause, the typhoid organism, and produced an immediate and easily diagnosed disease, typhoid fever. This made it possible to trace the contaminant to human wastes and to eliminate the problem in a fairly simple, clear-cut way. And, since people were threatened by a hazard usually created by their own wastes, it made obvious sense to solve that problem at public expense. People simply taxed themselves to build sewage treatment plants and to develop safe public water supplies. The problem was solved with relatively little controversy over who was at fault and who ought to pay the remedial cost.

In comparison, consider the typical modern water pollution problem of chemical wastes from industrial plants. The nature of this problem is also reported in Chapter 6. First, it is important to note that nearly all industrial chemicals are synthetic substances, most of which have been commercially used only since the 1940s. This means that the effects of these pollutants, whatever they turn out to be, are a new factor in environmental degradation. Second, the contaminants are not one or a few species of specific disease organisms, but hundreds of different chemicals, a relative few of which have been tested for their effects on health. Whereas typhoid contamination results in an immediate and easily recognized disease, some of the modern chemical contaminants are carcinogens and cause a disease—cancer—which may not develop until 10 to 20 years after a person has been exposed. Moreover, cancer is caused not by a single agent but by numerous chemicals, by irradiation, and probably by viruses. As a result, when a person develops cancer, no one can say whether it was due to drinking chemically contaminated water 10 or 20 years earlier or to some other cause. In addition, the effects of different chemicals may be more than additive; for example, two chemicals which do not themselves cause cancer may react in water (or, in the case of nitrate and amines, in a person's digestive tract) to produce a new substance which *does* cause cancer. All this shows that it is difficult to pin down the medical harm resulting from these new water pollutants, and that—unlike typhoid—one cannot focus on a single cause and then eliminate it.

On top of these technical difficulties, the elimination of modern water pollutants—again, in sharp contrast with typhoid—is complicated by economic, social, and political issues. As the evidence summarized in Chapter 6 shows, much of the chemical pollution of water supplies is due to the release of wastes from industrial plants. This means that today people are exposed to a hazard which is not of their own making. A city taking its drinking water from a stream polluted by industrial wastes could, at great expense, purify the water and remove the offensive chemicals. But, to pay for this, the people of that city would have to be willing to tax themselves for an environmental hazard created by someone else. Many people would regard this arrangement as unjust and would rather force the industrial plants to clean up their own wastes.

But that solution is not an easy one either. Waste controls cost money, and increased production costs reduce profits. From the industrialist's point of view, the people downstream are an economic threat if they insist on the cleanup. And if, under the pressure to control waste, the owners decide to close the plant, jobs are at stake and the problem becomes that much more complicated. Moreover, if a city or state sets stringent environmental laws to enforce control of industrial wastes, the affected industries can decide to move to another location in the United States, or even abroad, where the requirements are not so strict. On the other hand, if federal industrial waste standards are developed, some people will object to an extension of government controls over private enterprise.

This kind of comparison begins to give us a feeling for some of the reasons why water pollution, while at times less of an *immediate* threat to health than it once was, has become a more difficult problem to solve. The main reason is that water pollution, like nearly every other environmental issue, is deeply embedded in the fabric of economic, social, and political relations. There *is* an environmental crisis because we have not yet learned how, within these relationships, to take the steps needed to eliminate new and unprecedented assaults on our water resource.

Barry Commoner

Contents

Water Pollution

Assault on a Lake

August 2, 1885 was a bad day for Chicago. The people of the city were first knee-deep in water from a six-inch torrential downpour and later elbow-deep in raw sewage that poured out of their water taps. The two events were closely associated: Stormwaters had flushed out the sewers and carried sewage into Lake Michigan to the city's domestic water intake two miles from shore. Sucked into the intake, the sewage made the return trip to the city through the water supply pipes. The result was an epidemic of typhoid fever that lasted several years and that prompted a number of engineering improvements to break the sewage/drinking-water cycle.

These and subsequent improvements have counteracted waterborne diseases, but have not prevented the modern pollution difficulties that plague Chicago. The situation there illustrates serious shortcomings in current water pollution control strategy and theory which hinder our efforts to solve problems in Lake Michigan and elsewhere.

The Lake Michigan story facilitates examination of various interrelated factors of water pollution. Those factors now operate universally to degrade both fresh and salt water, but their very scope makes them difficult to put into perspective unless considered in the context of one circumscribed area. The degree of water pollution on the national and international scale will be discussed in Chapters 3 and 4.

Lake Michigan pollution became serious enough in the late 1960s to prompt Illinois Governor Otto Kerner to request an interstate conference under the provisions of the Federal Water Pollution Control Act. On January 31, 1968, the first session of the conference was held in Chicago. During

1968–1971, a series of 19 more sessions were held at which state and federal officials, scientists, businessmen, and conservationists discussed the problems of the lake. The objective was to formulate a series of recommendations to be endorsed by the federal government and written into regulations by the four states surrounding Lake Michigan (Wisconsin, Illinois, Indiana, and Michigan). The proceedings of these sessions appear in a 9000-page, 20-volume series which provides the basis for most of this chapter [1].

Conference speakers documented a long list of ills plaguing the lake, among them industrial pollutants including oils, phenols, chlorides, sulfates, polychlorinated biphenyls, phthalate plasticizers, and many other chemicals. The rising level of phosphate nutrients from municipal sewage and agricultural runoff already had produced overfertilization of inshore waters, thereby reducing the dissolved oxygen content in the water and causing widespread *eutrophic* problems. Conditions favorable to eutrophication were intensified in the summer of 1967 by phosphate nutrients from the decaying carcasses of five billion tons of dead alewives, small fish belonging to an ocean species that had been introduced into the Great Lakes by man. The alewives adapted only partially to the freshwater environment and thus were particularly vulnerable to natural temperature changes. Coho salmon, introduced into Lake Michigan to prey on the alewives and to provide sport fishing, had acquired very high levels of DDT which made many of them unfit for commercial sale [2]. Residue levels of DDT and its breakdown products were also high in other species of fish. These problems with fish, some experts concluded, were symptoms of serious biological changes brought about by the industrial, municipal,

Dead alewives on a Lake Michigan beach.

Lake Michigan's ecosystem was upset by the introduction of two ocean fish, the lamprey and the alewife, when the St. Lawrence Seaway opened a channel between the Atlantic Ocean and the Great Lakes. The lampreys decimated the lake trout. The alewife population, at first controlled by the trout, increased beyond proper proportions. Not well adapted to freshwater conditions, the alewives were vulnerable to the lake's comparatively rapid temperature changes. In the summer of 1967, dead alewives piled up on the beaches of the lake; their decaying carcasses added phosphates to the lake water, intensifying eutrophication. Subsequently, coho salmon, introduced to replace the trout for sport and commercial fishing and to control the alewives, became contaminated with DDT.

Coho salmon, ready to spawn, leap up a fish ladder which leads them into the hands of the Michigan Department of Natural Resources. Eggs are removed for the state's fish hatcheries, which later return a new generation of coho to the lake's tributaries.

and agricultural activities just mentioned, as well as the dumping of toxic dredging wastes and the discharge of hot water used for cooling in power plants into the lake, and pollution from harbor and shipping operations.

With data provided to him by the Department of Water and Sewers, Mayor Richard Daley summarized the Chicago situation at the first interstate conference session, January 31, 1968:

We know the problem of water pollution is increasing. During the past 20 years, the city of Chicago has had to increase the amount of chlorine used [as a disinfectant] to achieve water safety from 15 to 25 pounds per million gallons. This is an increase of 67 percent. We have had to increase the activated carbon dosage to remove objectionable odors and tastes from 24 to 41 pounds per million gallons, and other chemicals used to remove suspended matter from 103 to 119 pounds per million gallons. These are increases of 71 and 15 percent, respectively.

Symptomatic Treatment

A careful look at the individual ills besetting Lake Michigan—and representative of those plaguing water resources elsewhere—reveals certain conceptual flaws in water pollution control that, until they are corrected, make solutions seem remote.

Diversion. As an example of how these problems have developed, let us consider Chicago almost a century ago when it began to break the link between sewage and drinking water. The main countermeasure employed was something called *diversion.* This simply meant—and means today—routing sewage or other polluted material away from one area and into another. When the city was younger, the Chicago River had received Chicago's sanitary and industrial wastes, but unfortunately that river carried the wastes to Lake Michigan and dumped them where the city withdrew its drinking water. First, water intake tunnels were built two miles out into the lake. Then, in 1871, the Illinois and Michigan Canal was modified to force the Chicago River to carry its load of wastes the other way—into the Illinois hinterland. It seemed in the 1870s that sewage outlet and water intake had been safely separated. By 1885, however, increased drainage into the Chicago River had forced the river to return to its natural flow into Lake Michigan—a flow which became a disease-bearing flood in August of that year. To solve this problem, a canal was opened in 1900 joining the Chicago River with the Des Plaines River, which flowed into the Illinois River and then into the Mississippi River. As a result, the Chicago River, today bearing treated sewage and industrial wastes, flows westward into Illinois rather than eastward into Chicago's water supply.

All this seemed fine to Chicago residents, but Canadians and people living in other states bordering the Great Lakes feared their own water supplies might disappear down the Chicago River. As a result, the amount of water that Chicago could draw from Lake Michigan to force the river away from the city was limited by administrative and court decisions.

Despite parallel limitations today on the amount of Lake Michigan water that can be withdrawn elsewhere in Illinois, heavily populated suburbs north of Chicago (designated the North Shore Sanitary District) have been authorized to plan similar diversion schemes to keep their waste effluents out of the lake. The systems would divert discharges from a series of primary community waste treatment plants for further treatment elsewhere and then carry the waste water to the same drainage system that ultimately takes the treated Chicago wastes to the Illinois River.

All of this leads to the question: What happens to the wastes now being funneled into the Illinois and then into the interior river systems of the United States? Toxic materials in the wastes, such as chlorinated hydrocarbon pesticides and chemical insulators known as polychlorinated biphenyls (PCBs), are long-lived and will be carried great distances in the water.

Beyond this, diversion—as it has been applied over the year—has been extended by modern engineering techniques to include uses far beyond the emergency needs of nineteenth-century Chicago. The diversion concept has gained an established position that has yet to be fully assessed in light of modern pollution problems.

One of the most elaborate extensions of this concept was proposed recently by R. T. Jaske and W. A. Reardon of the Battelle Pacific Northwest Laboratories [3]. The investigators pointed out that a detailed study by Battelle scientists for the U.S. Atomic Energy Commission showed that the reduced flow of the Ohio River and, to a lesser extent, of the Mississippi during the late summer and early fall limited the number of nuclear power plants that could be built along these rivers, since the discharge of heated cooling water from the plants would not be diluted enough during these periods of low water to meet environmental stream temperature standards. Rather than installing expensive cooling towers or other systems to recycle cooling water, Jaske and Reardon proposed tapping the Great Lakes to cool the power plant condensers and at the same time to increase stream flow to dilute municipal sewage discharges. Citing the Chicago diversion as one historical precedent for their plan, the Battelle scientists pointed out that this elaborate scheme is within current technological capabilities since the engineering problems are similar to those dealt with in building the Grand Coulee Dam in Washington. They favored two sites on Lake Erie—one near the inland headwaters of the Allegheny River, the other at Cleveland. At each site, four 1500-megawatt nuclear power plants would be built. Water drawn from Lake Erie would cool the condensers and then be directed south, perhaps through reservoirs, to the Ohio River system to maintain an increased year-round flow in the river to accommodate other nuclear power plants. The eight nuclear plants would discharge up to 28,000 cubic feet per second into the Ohio River system. At Pittsburgh, where the Allegheny empties into the Ohio, the minimum year-round flow of the Ohio River would be some 25,000 cubic feet of water per second. This rate would be nearly twice the average low-water rate that now prevails during five months each year (June–October) [4].

As another feature, the authors suggested diverting Lake Superior's water to the Mississippi via the St. Croix River. This, coupled with the diversion of Lake Erie to the Ohio (which flows into the Mississippi), would not only assist the nuclear power industry but would also benefit another proposed diversion project in the lower Mississippi—the $10 billion Texas Water Plan to create a new water supply for agriculture in the west Texas highlands. The Texas Water Plan would require the diversion of some 15,000 to 22,000 cubic feet per

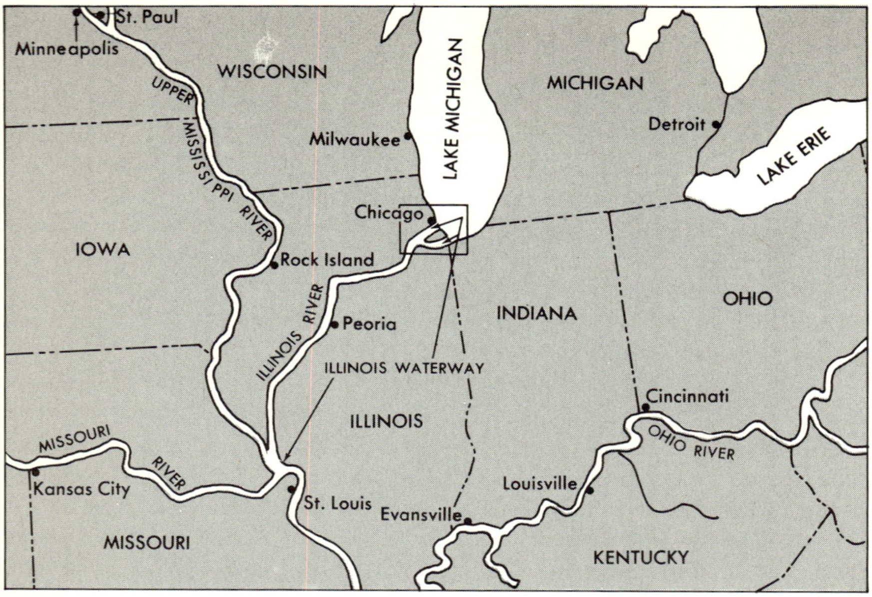

Water drawn from Lake Michigan receives Chicago wastes and is then diverted into the Illinois River. This keeps pollution out of the lake, but in spite of the treatment the waste receives before it passes into the river, it presents some problems downstream.

second of the Mississippi River by canal into Texas. Increased flow of the Mississippi achieved by diversion of the Great Lakes might insure that the water could be drawn off without lowering the level of the lower Mississippi and hence disrupting navigation.

Dilution. The Great Lakes diversion scheme was justified primarily in economic terms, with considerable weight given to savings in both municipal waste treatment costs and power plant equipment. Waste treatment savings would occur because the increased stream flow from diversion would promote *low-flow augmentation* and *dilution,* two other elements in the lexicon of conventional waste treatment theory. When stream flow is artificially increased or augmented, it can more effectively dilute polluted water; this increases the stream's capacity for dissolved oxygen and biological activity, with a consequent improved degradation of organic wastes [5] and a lessened need for investment in treatment facilities. However, many long-lived pollutants have ultimate biological effects no matter how thoroughly diluted; nutrients would survive to produce eutrophication and other long-lived pollutants requiring expensive advanced treatment for removal would simply be washed down the river.

Monumental hydrological, environmental, and ecological questions were not considered by the Great Lakes diversion planners, however. For example, sunlight might reach a greater proportion of bottom plant life in the already

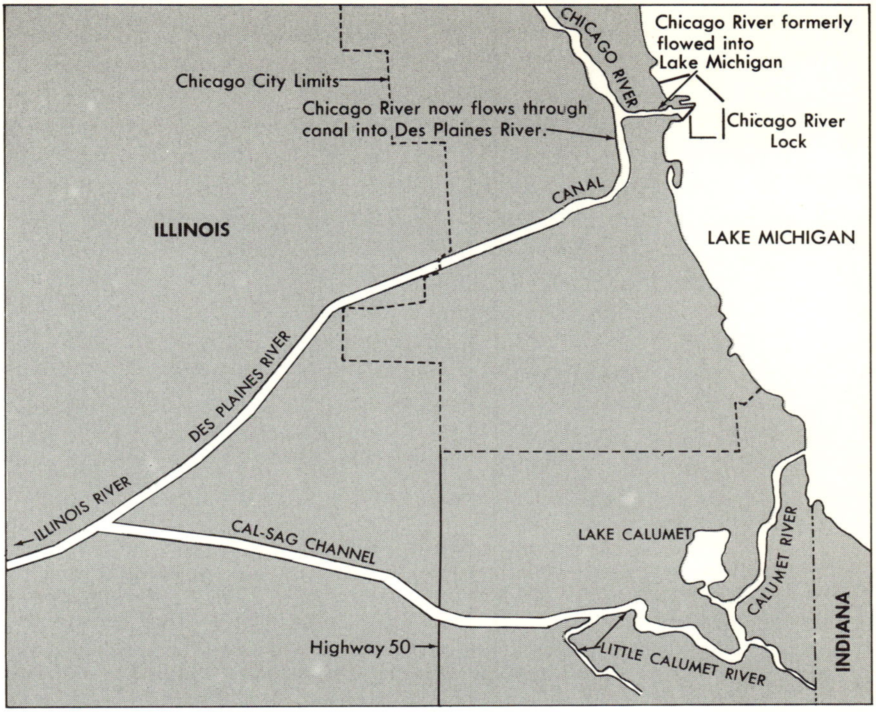

The Metropolitan Sanitary District of Greater Chicago. The area between Highway 50 and the Indiana border is a heavily industrialized city suburb.

shallow Lake Erie, enabling the vegetation to utilize the nutrients, and thus intensify eutrophic conditions. Canals and cooling water conduits might provide migration routes into the Ohio River for unwanted species of fish, just as other man-made structures have opened the way for undesirable species elsewhere in the Great Lakes.

The purpose in discussing the Great Lakes plan here is not to weigh its merits, other than to identify some of the questions that must be answered. Rather, the plan illustrates the way in which the techniques of diversion and augmented flow and the concept of dilution can be carried to extremes by modern technology.

Mixing Zone. Associated with the concept of dilution is that of the *mixing zone.* This term refers to lateral, vertical, and longitudinal forces, particularly those caused by turbulence, that disperse pollutants to a certain concentration within a given time and area in a stream or other body of water [6]. The mixing-zone concept is an important consideration in setting stream quality standards, which in the United States are generally determined by the allowable concentration of pollutants in a stream itself rather than in waste discharges. Formulas are used to calculate the anticipated mixing and dispersion of pollutants in a given stream. This enables officials to determine, for

Instead of limiting the amount of pollution discharged into a river or lake, dilution of the pollution is sometimes employed to meet standards requiring no more than a certain concentration of pollutants in a given amount of water. This means allowing more than the standard concentration in a mixing zone, where the pollution is diluted. The industrial effluent shown illustrates how waste mixes with more and more water and becomes less and less visible as it moves farther away from the plant. Formulas based on the turbulence of a given stream and other relevant factors are used to calculate the mixing zone.

example, whether waste discharges from a proposed industrial plant will raise pollution levels above the established standards. Although longitudinal mixing in particular can extend far downstream from the point of discharge, the area of most intense mixing is adjacent to the discharge pipe. This no man's land of high pollution is one price that must be paid for disposing of wastes in waterways, a problem stressed at the Lake Michigan conference session on March 23, 1971 by Donald I. Mount, Director of the National Water Quality Laboratory, Duluth, Minnesota:

> A mixing zone [as used for regulatory purposes] I feel by definition is a place where receiving water standards are not met. That is, the conditions are less than we are striving to achieve in the receiving water. . . . So the question you ultimately come to is how much sin is acceptable?

The answer depends upon who is being questioned. Mount's comment specifically concerned the degree to which heated cooling water from power plants and other industries should be permitted to warm Lake Michigan. Industrial representatives argued that there was no "sin," since heated effluents would do no meaningful harm to the lake. Investigators from the U.S. Fish and Wildlife Service, on the other hand, foresaw serious damage to aquatic life and an increase in eutrophication if anything but slightly heated effluents were permitted.

"Hot Water"

As of September 1970, one nuclear and 23 fossil-fuel power plants were operating at Lake Michigan. Five additional nuclear plants and one using fossil fuel were under construction and scheduled for operation by 1974; an additional nuclear plant has been proposed. The growth of industrial and municipal power needs, coupled with the inviting cooling waters of Lake Michigan, made it clear that sites for other new plants would be requested.

The installed generation capacity in 1968 on Lake Michigan was 7600 megawatts; the projected estimate for the year 2000 is 73,500 megawatts, of which 43,934 would be in nuclear plants. This estimate was made at the lake conferences by Yates M. Barber, Staff Assistant of the U. S. Bureau of Sport Fisheries and Wildlife, on the basis of projections developed in a Canadian study [7] and by the U.S. Federal Power Commission. Barber added that the output of waste heat at full capacity would be 414.67 billion BTUs per hour in the year 2000, nearly 14 times the BTU output in 1968. Furthermore, the total volume of water discharged from the plants in 2000 would be equal to a stream heated 20°F above the existing lake temperature and flowing into the lake at a rate of 91,000 cubic feet per second. This flow, over a year's time, would be equivalent to a sheet of hot water 4.6 feet deep covering the entire 14.3 million acres of the lake surface.

The effect of this heat on Lake Michigan was estimated by an interstate committee on thermal waste established during the Lake Michigan conferences. Allowing for loss of heat from the discharges to the atmosphere due to evaporation, convection, and radiation, the committee estimated that by 2023 generating power capacity would have increased enough to raise the annual average temperature of the entire volume of the lake 2°F, or from the 1969 average of 50° to an average of 52°. The magnitude of that increase (termed a conservative estimate by the committee) is clear if one compares it with natural temperature change in the lake. During the past 50 to 80 years, the temperature of the lake actually has declined by 2°, apparently from natural causes. The committee's projection means that, in about the same length of time, human activities could nullify the natural change by raising the average temperature 2°.

This prediction assumes that the heat from the discharges would be mixed evenly throughout the entire volume of Lake Michigan. Although theoretically possible, this would not happen under present circumstances since the heat discharge is limited generally to nearshore areas. The point is that man's activities now have the theoretical capacity to alter natural conditions in water on a scale previously reserved for nature.

Furthermore, the theoretical change predicted by the committee was particularly conservative since only heat input from power plants was considered. Large amounts of heat also are injected into Lake Michigan from industrial and municipal activities. Representatives of state and federal fish and wildlife services estimated at the conferences that about 25 percent of the total heat added to the lake by human activities was discharged from steel

The Zion Nuclear Power Station—two 1100 megawatt reactors—one of the six built or under construction on the shores of Lake Michigan. The water used by these and additional nuclear plants projected for construction by the year 2000 would be equal to a stream heated 20°F above the existing lake temperature and flowing into the lake at a rate of 91,000 cubic feet per second. While cooling towers can alleviate this problem, they are essentially another form of diversion—diverting the heat from the water, where it causes an immediate disruption of aquatic life, to the air, where it can produce local fogs. The long-term climatic problem of a continually increasing amount of waste heat remains the same, whether it is dispersed into the air or the water.

plants and municipal waste treatment facilities. Although the heat input from these two nonpower plant sources will continue to increase, it will be a proportionately smaller part of the total output of 430.77 billion BTUS per hour from power plants and other industrial sources that is projected for the year 2000 because of the far greater relative increase in heat input expected from power plants.

The official conferees from the federal government and the four states surrounding Lake Michigan concluded, on the basis of the evidence presented, that although warming throughout the lake as a whole would not occur, waste heat discharges from the increasing number of power plants might warm a large fraction of the inshore waters several degrees. Furthermore, the interaction of two or more thermal plumes might intensify biological effects more than anticipated. The conferees determined that a single large plant discharging into a spawning ground or across a migratory route could disrupt the balance of the affected species not just there, but throughout the lake. They also agreed that an unknown percentage of aquatic organisms such as adult fish, fish fry, and plankton would be killed by heat, physical damage, or chemical poisoning when they passed through the cooling systems of the power plants after being sucked into the intakes. The conferees could not agree on a solution, however. The Environmental Protection Agency (EPA) finally endorsed the position, proposed by the federal conferees and initially by conferees from three of the four states, that at a maximum distance of 1000 feet from the discharge of existing power plants or other large-scale industrial facilities, the receiving water temperature should be no more than 3°F above the natural temperature of the lake. In addition, the near-surface temperature 1000 feet from a discharge pipe should not exceed specified readings established for each month to prevent excessive heat during critical periods in the life cycles of aquatic organisms. As an additional requirement, large new power plants would have to build closed-cycle cooling systems to reduce discharges to a minimum. As of this writing, only Indiana has subsequently agreed to the federal proposals. EPA officials hope to resolve the differences in a reconvened conference this September (1973).

Even if the new controls are ultimately enforced, those requirements setting the temperature standard at a point 1000 feet from the point of discharge rely on the conventional practice of establishing a certain zone for wastes to disperse into water, despite findings that point to the danger of localized damage from the thermal pollution.

The cooling wastewater discharge from the seven existing and proposed nuclear power plants at Lake Michigan will contain radioactive wastes which also defy the mixing-zone concept. The high-level radioactive wastes will be shipped off to land-disposal sites, but low-level liquid wastes will be treated and eventually discharged in the cooling water. Some of the radionuclides injected into the lake will pose potential health threats, such as long-lived cesium 137 and the much shorter-lived iodine 131. Officials estimate that the total radioactivity of radionuclides thus discharged from the lake's seven power plants will be about one curie per year, exclusive of tritium—or radioactive hydrogen—which is a lesser health hazard and is not removed from the waste water. This estimate may be overly optimistic on the basis of experience at six relatively small nuclear power plants which went into operation elsewhere during 1961–1964. Four of the six plants discharged from 1.3 to 11.1 curies of radioactivity per year apiece [8]. Thus the much larger nuclear power plants at Lake Michigan would have to improve considerably on past performance in the nuclear power industry if the official prediction of a total of only one curie of radioactivity per year into the lake is to become reality.

Should a nuclear accident occur, discharges of course would be an altogether different matter. "One serious nuclear accident on the shore of Lake Michigan," said F. W. Kittrell of the U.S. Department of the Interior, at the second interstate conference session in Chicago, February 25, 1969, "although highly improbable, could release large quantities of long-lived radioactive material into Lake Michigan where it would be uncontrolled and could have very serious effects on future use of the lake."

Current federal regulations would permit an annual average of some 490,000 curies of radioactivity from routine plant operations if mixed throughout the entire volume of Lake Michigan, according to estimates by the Committee on Nuclear Power Plant Waste Disposal established during the conference. The one-curie-per-year level predicted for Lake Michigan power plant discharges, or even the much higher level suggested by past performance, therefore seems inconsequential, but the accumulation of radionuclides in the lake could still have serious effects. At the first conference session in Chicago, February 1, 1968, according to R. J. Schneider, Deputy Regional Director with the Federal Water Pollution Control Administration (FWPCA), now the Federal Water Quality Office: "The particular concern for Lake Michigan arises from the fact that it has a very small discharge rate, and any radioactive material entering into the lake will diminish only by natural decay. This may result in significantly increased levels of the longer-lived radioisotopes."

The threat may therefore be cumulative rather than immediate, since it would occur by concentration along the food chain—the same inexorable process that concentrates DDT from very small quantities in the water into much larger and more lethal quantities in higher animals. Treated radioactive wastes will be diluted in the cooling water discharge at carefully calculated rates to reduce radioactivity intensity below that permitted by U.S. Public Health Service drinking water standards. But biological magnification with respect to concentration will counteract the dilution process, making it possible for radioactivity to reach people by another route—to move up through the aquatic life that serves as the food chain for fish to those who eat fish from the lake. Although Mr. Kittrell stated that experience with wastes from smaller nuclear power plants now discharging to streams in Illinois and Wisconsin and to the lake in Michigan has not revealed such a problem, he warned that it must be evaluated when the larger plants begin to operate on Lake Michigan.

Evaluation *after* the contaminants have been released, however, could mean that the problem will become serious before it is officially acknowledged. This is illustrated by the pattern of damage from other long-lived contaminants, namely pesticides, in the lake.

Chemical Pollution

Even at the time DDT and its breakdown products were barely detectable in the waters of Lake Michigan, the chemical poison had accumulated to serious proportions in fish. This occurred after the pesticide had become concentrated in the lower forms of animal and plant life in the lake. Each successively higher animal then fed on more and more organisms containing pesticides; the food was metabolized and excreted, but the pesticides were retained. This led to a greater concentration of the chemicals in the higher animals than in the individual organisms those animals ate. Ultimately, the

Even when DDT and its breakdown products were barely detectable in the waters of Lake Michigan, the chemical poison had accumulated in fish and birds as it became more and more concentrated at successive levels in the food chain.

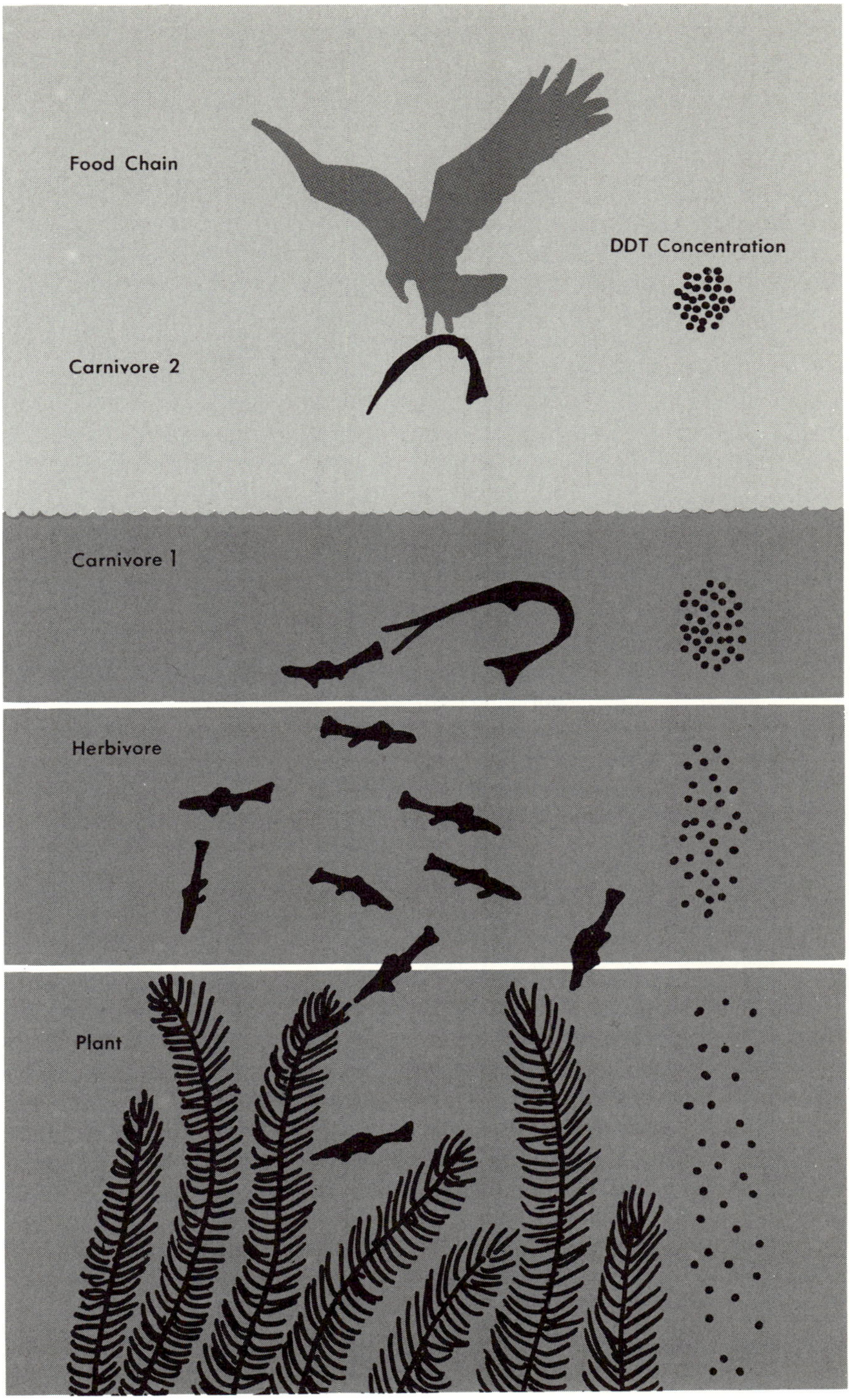
Food Chain
DDT Concentration
Carnivore 2
Carnivore 1
Herbivore
Plant

greatest concentration of the chemical occurred in the animals highest in the food chain.

An average of 19 parts per million (ppm) of DDT residues in canned Lake Michigan coho salmon prompted the Food and Drug Administration (FDA) to ban 32,000 pounds of the fish product from interstate commerce in 1969 [9]. The action was particularly discouraging to Michigan wildlife officials who had sponsored a seven million dollar program to stock the salmon in Lake Michigan to replace the depleted lake trout—a large fish commercially important in Lake Michigan until the late 1940s, when its numbers were decimated by the parasitic attack of sea lampreys that had worked their way up through the Great Lakes from the Atlantic Ocean.

DDT has not only made the coho salmon unfit for eating, but also has been implicated in the reduction of the salmon's breeding capacity—another chapter in the man-induced alterations of the biology of Lake Michigan. In 1968, coho fry tested by Michigan Department of Conservation scientists showed typical symptoms of pesticide poisoning, and larger than anticipated losses of young fish occurred in the hatcheries, according to Ralph A. MacMullen, director of that department. MacMullen told the February 6, 1968, session of the Lake Michigan conferences that:

> The concentrations are very close to those which have been shown to affect lake trout survival in some eastern states. And they actually exceed those which have been shown in British Columbia research to affect coho salmon under some conditions. We most certainly are close to the danger point. And we most certainly cannot stand a buildup.

Yet the buildup occurred, and it seems unlikely to drop sharply for some time, despite bans on most widescale uses of DDT instituted after 1968 in Michigan and Wisconsin. In 1970, for example, coho salmon tested by the Great Lakes Fishery Laboratory had average DDT concentrations of 14.1 ppm. The average for 19 lake trout was 19 ppm, with the range from 11 to 28 ppm. These and other tests showed no downward trend from concentrations in past years.

The 1970 Great Lakes Fishery Laboratory tests also revealed an average concentration of 0.27 ppm of dieldrin (a chemical cousin of DDT) in the lake trout and 0.12 ppm in the larger coho salmon. These averages exceeded the maximum value of 0.1 ppm for this pesticide in fish recommended by the Interstate Pesticide Committee formed under the auspices of the Lake Michigan conferences and they approached the FDA actionable level of 0.3 ppm; a number of individual fish did have dieldrin concentrations exceeding the 0.3 ppm level. Persistent levels of chlorinated hydrocarbon pesticides like DDT and dieldrin, despite closer regulation, may result from accumulations in 0430
older fish over a period of years, a factor that will be hard to assess until younger fish are examined in the future. The levels also reflect the presence in the environment of a widespread reservoir of durable chemicals that has built up since World War II and that will be with us for many years to come [10].

This chemical reservoir is the critical factor in the extent of long-term water damage, and local bans on pesticide use are only partial barriers to influxes from the reservoir. To give one illustration, Wisconsin investigators analyzed clams and other invertebrates near the mouths of rivers flowing into Lake Michigan and Lake Superior. In certain locations, they discovered somewhat high levels of dieldrin and sought out the sources of the pesticide upstream. One was a wool-dying operation where dieldrin was used to mothproof the wool; some of the chemical was washed into the plant's sewage system and subsequently into a stream. Once plant officials stopped using dieldrin for mothproofing, everyone thought all would be well, but new tests indicated that lesser amounts of dieldrin were still present in the plant's effluent. Puzzled investigators finally tested the virgin wool shipped into the plant and discovered that it, too, contained traces of dieldrin. They concluded that the wool had become contaminated even before it was sheared from the sheep's back, probably when the range animal was routinely treated with a disinfectant mixture containing dieldrin.

An unwitting addition to the reservoir of dangerous, long-lived materials in man's environment has been the accidental release of another chlorinated hydrocarbon called polychlorinated biphenyls (PCBs). Utilized for a number of industrial purposes, this material is now widespread in the biosphere, like DDT, has become concentrated along the food chain [11]. In his report on the 1970 pesticide assays in Lake Michigan, Dr. Mount pointed out that lake water samples examined in one laboratory contained 50 to 75 parts per trillion of PCBs. By comparison, studies in the late 1960s by the Michigan Water Resources Commission indicated that the concentration of DDT was nearly always less than 10 parts per trillion in open lake water.

Concentrations of PCBs along the food chain already have been found in marine life, sea birds, and birds of prey [12]. The hazards of PCB are not yet completely understood and may be complicated by the fact that some PCB mixtures, especially those manufactured in Europe and Japan, appear to be contaminated with trace amounts of toxic impurities. Although concentrations in the environment are much too small to be a direct threat to human health, PCB levels of only a few parts per billion (ppb) are toxic to some marine crustaceans, mollusks, and fish. The chemicals have affected the reproductive system in some birds, although apparently not as markedly as has DDT, and laboratory tests have shown that concentrations as low as 10 to 25 ppb inhibit the growth of two types of one-celled algae found in the ocean [13]. Environmental hazards are increased by interactions between PCBs, DDT, and other chlorinated pesticides which produce an effect that is greater than that caused by one of the chemicals alone [14].

Other chemicals called phthalates are also accumulating in Lake Michigan. These substances are widely used as plasticizers to impart flexibility to polyvinyl plastics; they also have been used in insecticides [15]. Even though chemically different from DDT, the capacity of phthalates to become concentrated along the food chain in the same fashion as DDT has been established at the Fish-Pesticide Research Laboratory in Columbia, Missouri. Although the chemicals were of fairly low direct toxicity to fish, in tests laboratory levels as low as 3 ppb reduced the growth and reproduction rates of certain crustaceans. Foster Mayer, Jr., and colleagues at the laboratory wrote:

> Although phthalate esters cannot be impugned as an environmental hazard, their subtle effects on aquatic organisms resulting from long-term, subacute exposures is unknown . . . there are indications that these compounds can be detrimental to aquatic organisms at low chronic concentrations.[16]

The biological concentration of these various long-lived contaminants defies the concept of dilution upon which much of waste disposal engineering rests. Thus, cesium 137 discharged from the Zion Nuclear Power Station north of Chicago, although diluted in the cooling water discharge, will be concentrated in vegetation watered by Lake Michigan and further concentrated in the livestock feeding on that vegetation. Minute quantities of cesium 137 in the water will be concentrated by some 9500 times in the bodies of sunfish and by some 3000 times in the bodies of carp [17]. In any event, the radioactive potential of the material will be passed on to man at some point in the food chain even though the level of radioactivity in his drinking water is kept very low. Immediate damage to the human body from ionizing radiation received in this way is probably not a threat; cumulative damage—such as the inducement of cancer over a long period of time or genetic effects that will affect some human offspring in one way or another—is a possibility that must be weighed against the benefits of nuclear power. With regard to genetic effects, of radiation, the authors of the definitive book *Water Quality Criteria,* prepared for the California Water Quality Control Board, said:

> For genetic effects, however, there appears to be no possible threshold dose below which some effect may not occur. Any amount of radiation, no matter how small, is bound to produce some mutations. Once completed, mutations are irreversible and almost all mutations are deleterious. Irradiation of a population from any source is bound to have some genetic consequences. [18]

Problems Accumulate

Although the potential damage to human health from radioactivity, PCBs, phthalates, and a host of organic and inorganic chemicals that are accumulating in the environment is impossible to quantify, the history of durable, man-made pollutants reveals a multitude of ways in which nature reincorporates our wastes back into living systems—for better or for worse. The complexity of this biological process was far from the minds of the sanitary engineers desperately struggling to devise a method to keep disease-bearing excrement out of Chicago's drinking water in 1885. Yet their early patterns of thinking about waste disposal persist today, with merely a token integration of the insights provided by the biological sciences. Often enough, it is only when the damage has been done that biological investigation is instituted, and by then it is already too late; witness the insecticide decimation of bird populations or, even more seriously, radiation-induced genetic changes in human beings.

The most visible evidence of this shortsightedness is the buildup of heavy algal scums that stain nearshore waters in U.S. streams and lakes, Lake Michigan included. The algae proliferate by consuming nutrients, including the phosphates and nitrates man generously supplies. One of the largest sources of this supply of algal food is in the discharge of secondary sewage treatment plants. In nature, inorganic nutrients produced by bacteria are intricately balanced by a number of factors. The secondary sewage treatment plant, on the other hand, releases nutrients in uncontrolled amounts that feed the algae and consequently produce the condition known as eutrophication that reduces the water's dissolved oxygen content.

The visible decline of Lake Michigan due to eutrophication was compared with the condition of Lake Erie by A. F. Bartsch of the Federal Water Pollution Control Administration (now the Federal Water Quality Office) at the February 1, 1968 session of the Lake Michigan conferences:

> Lake Erie at some point in the past. With increasing time, nutrient levels will increase until finally the entire lake becomes involved. With certain reservations, Lake Erie can be viewed as a prototype and a preview of what can happen in Lake Michigan if nutrient-bearing waste input continues unabated.

Data available since 1968 show that the waste input has indeed continued. In a series of reports to the Lake Michigan conferences, James C. Vaughn, Engineer of Water Purification at Chicago's Department of Water and Sewers, traced the steady buildup of phosphate concentrations in Lake Michigan water near the Chicago shoreline and in open water along the Illinois shoreline. Four-month average concentrations of phosphates in mixed offshore and shore water at the Chicago South Water Filtration Plant increased from a 1966 range of .024–.042 to a 1971 range of .061–.086 milligrams of total phosphates per liter of water.* In other words, the quarterly average concentrations rose, with some fluctuations, to values in 1971 that were two to three times those in 1966.

The annual average for total phosphates in 1969 was .05 mg/l; in 1970, the average was .06 mg/l, three times the permissible annual average value set by the Illinois Pollution Control Board. The control board value, an annual average of .02 mg/l of phosphates of water, is considered to be the level above which excessive algal growth will be likely.

Reports from Mr. Vaughn demonstrate a similar trend in open water. In 1970, annual averages along a 12-mile stretch of water from Chicago up to Waukegan exceeded .09 mg/l of phosphates at four of ten sample points one to five miles from shore. These averages, then, were more than four times the permissible level set by the Illinois Pollution Control Board. In fact, the 1970 average was below that value at only one of the ten test points, and that one was five miles offshore. In 1971, the permissible value was exceeded at 20 of 21 sample points.

---

*Concentrations of contaminants in water and Public Health standards for permissible concentrations are usually given in milligrams per liter (mg/l), but are sometimes expressed in parts per million (ppm). Since pure water weighs one thousand grams per liter, or one million milligrams, one mg/l is comparable to one ppm.

Steady increases in the phosphate level were associated with rising concentrations of microorganisms over the past 20 years, as measured in mixed offshore and shore water at the Chicago South Water Filtration Plant. These organisms have caused a number of expensive problems at the plant—as they have at water treatment plants elsewhere. For example, the filamentous algae *Cladophora* proliferated on the bottom of Lake Michigan under the influence of waste nutrients and broke loose late in the summers of 1968–1971, clogging intake screens to the point that they had to be cleaned by hand. Two other microorganisms present in the winter recently have affected the water chemistry to the degree that greater doses of coagulants are now needed to reduce turbidity in the drinking water supply; a third microorganism imparted a cod-liver-oil taste and smell to the water, necessitating a counteractive increase in chlorination or activated carbon treatment. Algal interference with water treatment also has occurred at facilities in Green Bay, Sheboygan, Milwaukee, Waukegan, Evanston, Gary, Michigan City, Benton Harbor, Holland, Grand Rapids, and Muskegon.

On the basis of evidence produced at the Lake Michigan conferences, the four surrounding states have undertaken a program to improve municipal waste treatment to achieve removal of 80 percent of the total phosphorus in waste treatment plants that border the lake. Industrial wastes will require separate treatment if not treated in the municipal plants. The effect of this action on the lake's ecology remains to be seen; the program will not produce substantial results immediately, since initial time has to be devoted to financing, personnel training, and alteration of facilities.

The prognosis still appears serious for Lake Michigan for several reasons. In the first place, removing even 80 percent of the phosphates in man-made wastes leaves quite a lot of food for the algae. The latest estimates of phosphorus input into the Lake Michigan Basin made by the Phosphate Technical Committee are: A total of about 16.7 million pounds of phosphorus compounds—a conservative estimate—enters the basin each year. Of that total, about 13.3 million pounds come from municipal treatment plants, other domestic sources, or industrial plants; two million pounds are in runoff from agricultural lands, urban and rural areas, and land development or construction areas; 1.4 million pounds are contributed by miscellaneous sources such as decaying leaves and rotting crop vegetation.

In addition to the anticipated 80 percent reduction of phosphates in municipal and industrial discharges, encouragement of improved agricultural practices should cut phosphate contributions from farm lands. For the moment, assume that this gives us a future across-the-board reduction of 80 percent—a very optimistic assumption, according to EPA experts. Under the best of circumstances, this still leaves us with some 3.3 million pounds of phosphorus compound input each year. This in itself is a substantial amount of nutrition for aquatic life, and it will be magnified in coming years as both population and industry grow. Some forecasts for the Lake Michigan Basin predict population growth will rise from its 1960 level of 4.2 million to 12.1 million by the year 2020, and industrial wastes in the same period will increase by a factor of 3. Nutrients, such as nitrogen and phosphorus, may be expected to show somewhat the same growth pattern.

Also, the future of eutrophication in the lake depends in large part on whether inorganic nutrients continue to accumulate. Lake Michigan is a huge cul-de-sac which is flushed out very slowly through the Straits of Mackinac into Lake Huron. So slow is the flushing that an estimated 95 percent of the annual input of phosphates into the lake are retained there, according to A. F. Bartsch (FWPCA). As an essential element in plant and animal growth, phosphorus tends to be retained in the natural aquatic cycle—now in living tissue, now in decaying organic matter, now in inorganic phosphates excreted by various organisms. Its retention time in the lake therefore is even greater than that of a stable dissolved pollutant that follows the main flow of the lake water. Because of the factors of wind, current, shoreline barriers, and thermal stratification, and the man-induced factor of intensified pollution in shallow waters, it might take 1000 years for the lake to flush itself of 90 percent of the general pollution load even with very effective pollution controls, according to an estimate made at the February 1, 1968, session by Donald J. Baumgartner, Chief of Oceanography at the Pacific Northwest Water Research Laboratory, Corvallis, Oregon. Wastes like phosphorus that build up in the aquatic system may follow a flushing pattern that covers an even longer span of time.

Furthermore, although phosphorus in the form of inorganic phosphates appears to be the most critical nutrient in eutrophication, other nutrients—particularly nitrogen—also contribute to the problem. Phosphates are more easily removed in the waste treatment process and nitrogen is naturally more abundant, so phosphorus reduction in treatment plants more effectively reduces eutrophication. Nevertheless, very large amounts of nitrogen in inorganic nitrates are released in water due to human activities. In Lake Michigan, A. F. Bartsch estimated the annual input of nitrogen from all sources to be about 166 million pounds, of which 81 percent is retained in the basin each year.

Other trends in lake chemistry point to Lake Michigan's deterioration, trends which go hand in hand with the decline in water quality in rivers and lakes elsewhere. From 1870 to 1967, concentrations of sulfates, chlorides, and total dissolved solids (total content of dissolved carbonates, bicarbonates, chlorides, sulfates, phosphates, and other substances) rose steadily, as depicted in the following graphs.

These trends were reported by W. F. Carbine, then Regional Director, Great Lakes-Central Region, U.S. Bureau of Commercial Fisheries, who said these three substances were the only ones for which reliable, standardized measurements could be obtained over a long period of time. Nevertheless, Carbine saw these trends as broadly representative of what is happening to the concentrations of many (though by no means all) other chemical substances and similar to the steady increases that took place in Lake Erie prior to the more recent sharp increases there.

The rising curve for chlorides was singled out as particularly significant. Chloride concentration increased only 2 ppm during 1905–1955, a period of

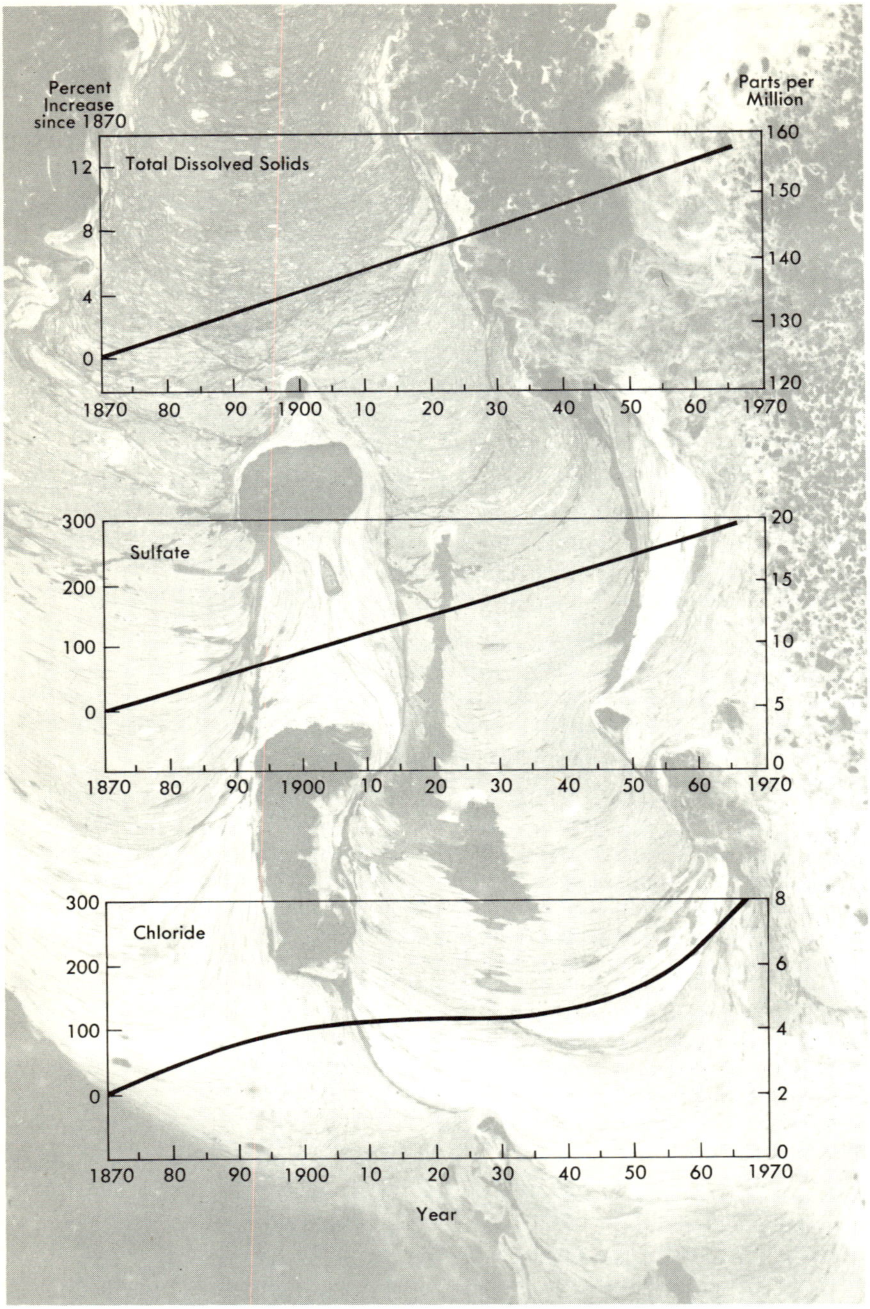

The rising concentration of chemical substances is one measure of the decline in water quality, and is typified in this chart of the rise in chloride, sulfate, and total dissolved solids in Lake Michigan since 1870.

50 years, but then rose another 2 ppm in only one-quarter of that time during the 12-year period from 1955–1967. This sudden, accelerated rise was similar to the trend in Lake Erie as eutrophication intensified in that body of water.

Even though estimates were that Michigan salt and brine industries injected at least half of the average daily chloride input into Lake Michigan in 1971, the broader importance of the rising level of various dissolved materials was that it was indicative of ever-increasing amounts of many different types of pollutants from activities such as steelmaking, food processing, chemical manufacturing, paper making, and petroleum processing. The industrial problems at Lake Michigan can be found throughout America and other industrialized lands.

Pollutants reach Lake Michigan in air pollution as well as in liquid wastes. Prevailing westerly winds carry pollutants, discharged from industrial smokestacks at the south end of the lake, over the water where they can be deposited as dry fallout or in rainfall. The degree to which this may happen was estimated recently for metals by John W. Winchester and Gordon D. Nifong, both at the University of Michigan, Ann Arbor [19]. On the basis of production figures, published emissions measurements, and other information, the authors calculated that significantly large amounts of iron (95,000 tons), manganese (5100 tons), copper (3500 tons), and zinc (4300 tons) are discharged into the atmosphere each year by Chicago, Milwaukee, and northwest Indiana businesses, primarily the iron and steel industries. Large amounts of titanium (2600 tons) and nickel (1100 tons) are released into the air annually by coal, coke, and fuel oil burned for heating, manufacturing, and generation of electricity. Lead (2400 pounds per year) is in the exhaust from gasoline-burning engines and, to a lesser degree, in smoke from other fuel.

On the basis of the strength and direction of prevailing winds and previous studies of the rate at which airborne particles are deposited in water, the investigators concluded that some ten percent of the overall metal pollution would end up in Lake Michigan, and part of what was left in the air undoubtedly would travel to other bodies of water. The scientists assumed that an increase in the concentration of any element in the water by as much as 0.1 mg/l (about the same as 0.1 ppb) was potentially significant to the ecology of the lake. Allowing for natural input of metals into the lake and for natural flushing of them from the lake, a number of metals from the air pollution were building up in the water in important amounts—among them copper, nickel, and lead, each of which could have detrimental effects on the lake's ecosystem. In addition to the metals studied by Winchester and his colleagues, other air pollutants ranging from chemicals such as pesticides to asbestos fibers from construction projects find their way into the water. Even sulfur dioxide, a gas product of combustion, is changed to an acid pollutant in the atmosphere and washed down in rain [20].

Still other sources of pollution plague Lake Michigan. Harbors fill with sediment from rivers to the point that dredging is necessary to maintain adequate depths for navigation. In 1970, 30 harbors were scheduled for dredging by the U.S. Army Corps of Engineers. The dredging wastes were considered partially polluted with industrial and domestic wastes in 11 of these harbors. Scheduled operations called for dumping into the open waters of the lake 322,000 cubic yards of polluted wastes—enough to cover one city block to a 40-foot depth. Another 2,330,000 cubic yards of polluted material would be dumped in confined areas as part of a new program to limit dredging pollution. Some 1.8 million cubic yards of unpolluted dredging wastes also were scheduled for dumping into the open lake. The ecological consequences of such dredging projects are unclear, but polluted wastes have been shown to have serious effects in dredging disposal sites off New York City (this will be discussed elsewhere).

Another problem is that decreasing industrial discharges into Lake Michigan does not mean that the wastes will disappear. Some waste material is now dumped into landfills which, if poorly designed, allow heavily contaminated liquids to leak out into surface waters. Reliance on deep wells for

In an effort to ease the lake's burden of industrial waste, another form of diversion has been used: Some waste material is dumped into landfills which, if poorly designed, allow contaminated liquids to leak out into surface waters. Other waste materials are injected into deep wells which may pollute ground water. Deep wells have problems of their own.

Problems of deep-well disposal of liquid waste:

1. Improper pressure monitoring, resulting in failure of injection equipment or fracturing of disposal formation.
2. Mining, excavations, or additional wells might lower subsurface pressures. This could trigger escape of waste fluids from the disposal zone (into other formations, water reservoirs, or to the surface).
3. Wastes under pressure can escape through abandoned and improperly plugged oil, gas, or water wells nearby.
4. Corrosion of well with age, allowing fluids under pressure to escape.
5. Natural or induced fractures in rock formations through which wastes may escape.
6. Corroded or broken casing through which wastes can escape.
7. Corrosion break in injection tubing, allowing wastes to enter space between tubing and casing (annulus).
8. Poor cement job. Wastes under pressure can break out between outer casing and the surrounding geological formation if outer casing is not adequately cemented to formation.
9. Leaking or faulty packer, allowing wastes to enter annulus.
10. Several wastes injected together can react and form new compuunds. Wastes can also react with the injection formation, or with other wastes injected into nearby wells which share the same disposal area. Such reactions could create toxic fluids or poisonous gases in some cases.
11. Fluid could invade injection zone by means of osmosis, with resulting rise in pressure and fracturing of zone.
12. Clogging of injection tubing or disposal formation can cause pressures to rise abruptly, with consequent possibility of rupturing of tubing or fracturing of formation and consequent escape of wastes.
13. Wastes under pressure can reduce friction between rocks under stress, triggering earthquakes.

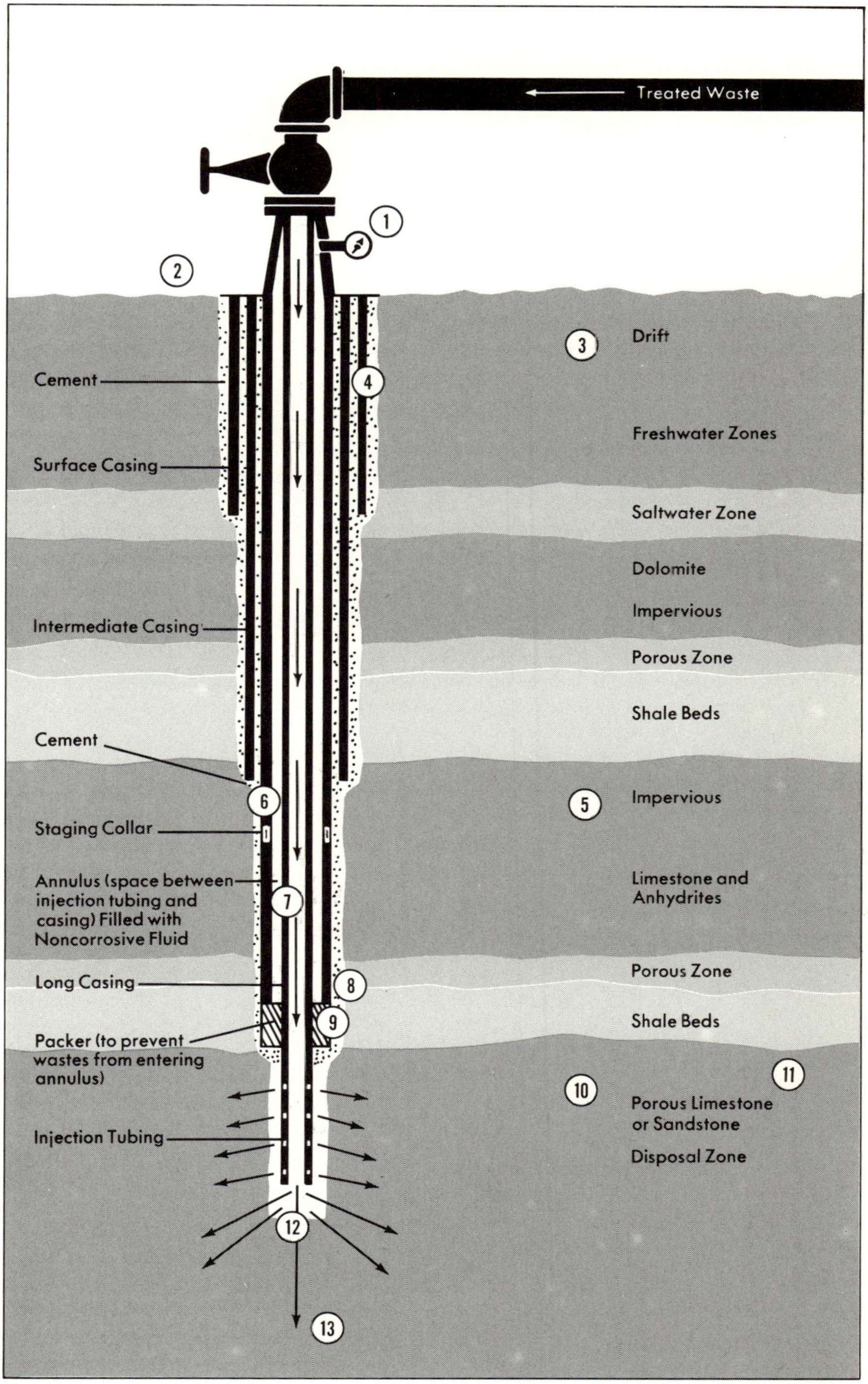
Treated Waste
1
2
3
Drift
Cement
4
Freshwater Zones
Surface Casing
Saltwater Zone
Dolomite
Impervious
Intermediate Casing
Porous Zone
Shale Beds
Cement
6
5
Impervious
Staging Collar
Annulus (space between injection tubing and casing) Filled with Noncorrosive Fluid
7
Limestone and Anhydrites
Porous Zone
Long Casing
8
9
Shale Beds
Packer (to prevent wastes from entering annulus)
11
10
Porous Limestone or Sandstone
Injection Tubing
Disposal Zone
12
13

Chicago's Lake Shore in the late 1880s. The city's water pollution problem then was a too-close association of sewage outfall into Lake Michigan with drinking water intake from the lake. The result was an epidemic of typhoid fever. Photograph courtesy of Chicago Historical Society.

waste disposal involves the risk of contaminating subsurface waters that are critical to future domestic needs and in which pollutants persist even longer than in surface waters. At Lake Michigan, the deep-well disposal technique is being used by steel companies at several locations. In 1970, workers at the U.S. Steel plant in Waukegan, Illinois, hauled 30,000 gallons of acid wastes daily to another plant in Gary, Indiana, for deep-well injection. One deep-well disposal system in Illinois also received steel mill wastes; another, oil field wastes.

Thus, the decline of Lake Michigan and of lakes and rivers elsewhere has been brought about by direct contamination and by indirect pollution in fallout from the atmosphere and in subsurface waters that eventually find their way to ground level. At the heart of the water pollution crisis is the same narrow view of water pollution "control" that produced such concepts as the mixing zone, rate of dilution, diversion of waste streams, low-flow augmentation, and the absorptive capacity of streams. The oversight is that wastes are discharged not just into water but rather into a gigantic, worldwide hydrological cycle in which water circles endlessly in different physical forms from land, to lakes and oceans, to the atmosphere, and back to land again. A pollutant injected at one point in the cycle may be transferred to other portions, thus poisoning large segments of the cycle—a process that has already occurred with long-lived chemicals such as DDT and PCBs. For such pollutants,

Chicago's Lake Shore in 1970. The pollution of Lake Michigan that appeared in the 1880s to be a one-dimensional problem with a simple sanitary engineering solution has now become far more complex as the city and its industry have grown and changed. Photograph courtesy of Chicago Historical Society.

the entire hydrological cycle rather than a localized area becomes, in effect, the mixing zone. The capacity of the various forms of water to transport durable pollutants in solution, in suspension, or in a vaporized state defies the concept of safe dilution of wastes. Furthermore, the very processes of dissolution, oxidation, and biodegradation that we rely on to make wastes harmless in water may, in reality, sometimes generate a compound far more biologically harmful than the original substance; this occurs when the metal mercury, which itself passes without serious incident through man's gastrointestinal tract, is changed by sedimentary microorganisms into methylmercury—an organic compound that is absorbed through the human gut and that can cause irreversible brain damage [21]. (Biological and chemical interactions will be discussed in the next chapter.)

Such complicated microbial action dramatizes another dimension of our shortsightedness: The vast hydrological cycle is essential to all life and, as such, is so tightly linked to the existence of plants and animals at every level of biological complexity as to be inseparable from it. Our imperfect understanding of this relationship is illustrated by the use of biodegradation in secondary waste treatment facilities to decompose organic material without consideration for the fact that the products of that decomposition become a feast to organisms elsewhere, producing overgrowths that already have degraded Lake Erie, Lake Michigan, and many other waterways. Such degradation makes domestic use of water increasingly difficult, as we have seen from the experience at the Chicago water purification plant. Of even more immediate danger to human health is the biological magnification of persistent pollutants along the food chain until they even reach man's table in meat or vegetables. This magnification, perhaps more than any other factor, demonstrates the fallacy of man's reliance on the "harmless" dilution of his wastes. In fact, many waterborne wastes will be concentrated in living systems rather than dispersed—a fact of life that did not matter much back in 1885, but that is of vital importance in today's industrial societies.

This leads to the conclusion that a modern science of water pollution must be consistent with the characteristics of the all-pervasive water cycle in the physical and biological world, rather than dependent on the more limited, often pragmatic, concepts developed in the early days of sanitary and water purification engineering. As we have seen, both life and industry are each day more dependent on water as a natural resource. As hard as we try, however, we cannot separate the consequences of these often conflicting users, since the hydrological cycle links the effects of all water use. The most important use of water has been to support basic life processes, and that will continue to be so despite our need to put water to other uses. To learn how we can use water for other purposes without destroying its primary life-support function, it is necessary to look more deeply into the characteristics of water itself. Those characteristics tell us much about what to expect from the pollution now contaminating the world's biosphere.

[1] *Conference on the Matter of Pollution of Lake Michigan and Its Tributary Basins*, Sessions 1–3, Environmental Protection Agency (Washington, D.C.: U.S. Government Printing Office), 9167 pages, January 31, 1968 to March 25, 1971 (available at various locations, including libraries designated as official repositories of federal documents).

[2] Harmon Henkin, "Problems in PPM," *Environment*, May 1969, Vol. 11, No. 4, 24ff.

[3] R. T. Jaske and W. A. Reardon, "Low-Flow Augmentation of the Ohio and Upper Mississippi Basin as a Means of Improved Planning for Water Resources Utilization," 15th Conference on Great Lakes Research, Madison, Wisconsin, April 7, 1972.

[4] *ORSANCO 1970*, 22nd Yearbook of the Ohio River Valley Water Sanitation Commission, 414 Walnut St., Cincinnati, Ohio 45202, 1971, 36 pages.

[5] Karl Imhoff, W. J. Müller, and D. K. B. Thistelthwayte, *Disposal of Sewage and Other Waterborne Wastes*, Second Edition (Ann Arbor: Ann Arbor Science Publishers, 1971), 405 pages.

[6] Jack Edward McKee and Harold W. Wolf (eds.), *Water Quality Criteria,* Second Edition (Sacramento, California: State Water Quality Control Board, 1963).

[7] H. G. Acres, *Thermal Inputs to the Great Lakes 1968–2000,* Canada Centre for Inland Waters, Niagara Falls, Ontario, 1970, 93 pages.

[8] J. O. Blomeke and F. E. Harrington, "Waste Management at Nuclear Power Stations," *Nuclear Safety,* May-June 1968, Vol. 9, No. 3, 239–48.

[9] Harmon Henkin, "Problems in PPM," *Environment,* May 1969, Vol. 11, No. 4, 24*ff.*

[10] George M. Woodwell, Paul P. Craig, and Horton A. Johnson, "DDT in the Biosphere: Where Does It Go?," *Science,* December 10, 1971, Vol. 174, 1101.

[11] Robert Risebrough, with Virginia Brodine, "More Letters in the Wind," *Environment,* January–February 1970, Vol. 12, No. 1, 16*ff.*

[12] S. Jensen et al., "DDT and PCB in Marine Animals from Swedish Waters," *Nature,* October 18, 1969, Vol. 224, No. 5216, 247–50.

[13] Jerry L. Mosser et al., "Polychlorinated Biphenyls: Toxicity to Certain Phytoplankters," *Science,* January 14, 1972, Vol. 175, 191–92.

[14] Allen L. Hammond, "Chemical Pollution: Polychlorinated Biphenyls," *Science,* January 14, 1972, Vol. 175, 155–56.

[15] Kevin P. Shea, "The New-Car Smell," *Environment,* October 1971, Vol. 13, No. 8, 2–9.

[16] Foster L. Mayer, Jr., David L. Stalling, and James L. Johnson, "Phthalate Esters: An Environmental Contaminant," Midwest Regional American Chemical Society Meeting, St. Louis, Missouri, October 28–29, 1971.

[17] McKee and Wolf, *Water Quality Criteria,* p. 353 (see [6]).

[18] Ibid, pp. 345–46.

[19] John W. Winchester and Gordon D. Nifong, "Water Pollution in Lake Michigan by Trace Elements from Pollution Aerosol Fallout," *Water, Air, and Soil Pollution,* 1971, Vol. 1, 50–64.

[20] Gene E. Likens, Herbert F. Bormann, and Noye M. Johnson, "Acid Rain," *Environment,* March 1972, Vol. 14, No. 2, 33–40.

[21] John M. Wood, "A Progress Report on Mercury," *Environment,* January–February 1972, Vol. 14, No. 1, 33–39.

An Uncommon Friend

Without water, this would have been a vastly different planet—and a lifeless one. In its beginning, the earth has been described as a hot mass of molten substances. As it cooled, the heaviest elements, such as iron and nickel, formed a molten core in the earth's interior; the lighter elements hardened and formed the earth's crust.

For a time the earth is thought to have been completely covered with clouds of water vapor and then to have cooled enough to allow condensation of the vapor into water droplets. Rain fell, resulting in the formation of large, freshwater seas. The clouds parted, letting in vital energy from the sun, and ever since that time water has played a major role in shaping the face of the earth and in the evolution and maintenance of life [1].

Water constantly circulates from one part of the earth to another, sometimes as water vapor in the atmosphere, sometimes as rain and snow falling on the land and the oceans, sometimes as rivers, streams, and ground water flowing over and under the earth's surface, sometimes held for long periods in lakes or for longer periods in oceans, sometimes frozen for a day, a season, or a millennium. Water moves in what is essentially a closed cycle, the *hydrological cycle* (pages 32–33), driven by solar energy. The oceans and seas contain 97 percent of the earth's circulating water and most of the remainder is held in the polar icecaps. Less than one percent is distributed as fresh water in streams and lakes, ground water, and water vapor in the atmosphere. Yet this is no small amount. Solar energy evaporates 80,000 cubic miles of water from the oceans and 15,000 cubic miles from land surfaces and lakes annually; and annually 24,000 cubic miles of water return to the land as rain and snow (enough water to cover Texas to a depth of 475 feet) [2]. The amount of water

evaporated from any particular part of the earth's surface depends on temperatures, latitudes, wind velocities, and other physical factors. Areas close to the equator with their high temperatures, for example, have higher rates of evaporation than areas farther north or south. But the climatic patterns which help to determine the rate of evaporation are themselves determined in part by the movement of water vapor through the atmosphere from one part of the globe to another.

The movement of the water that is evaporated from the land and the oceans is an important factor in the redistribution of heat around the globe. (Heat is also distributed by ocean currents.) Heat is absorbed when water evaporates, and the atmosphere becomes a reservoir of heat energy. This heat energy then drives the hydrological cycle which generates the atmospheric processes involved in weather and climate [3].

Water also moves over land and in subsurface areas. About one-third of the annual precipitation in the United States reaches surface waters—brooks, streams, and rivers. Some of the water that penetrates the ground for a short distance is drawn back up to the surface and evaporates, but a large portion of this water remains stored in the ground. Subsurface water constantly flows downward in response to gravity, filling all areas that contain porous materials (gravel, sand, sandstone, clay) until it reaches impermeable rock. Eventually subsurface water may emerge again as springs or may be tapped by wells.

The intimate interrelationships of water with the atmosphere and with all forms of life can be dramatically illustrated by the now worldwide distribution of DDT [4]. DDT has been found in organisms in areas as far removed from centers of agricultural activity as Antarctica. The abundant nutrients in

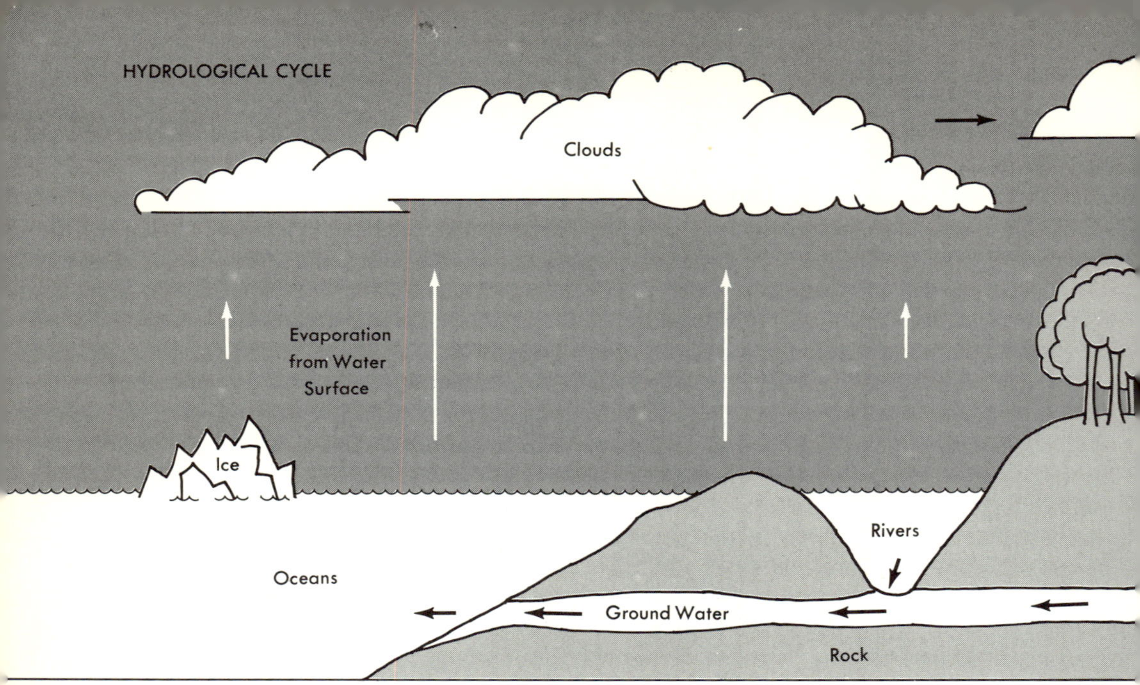

the offshore waters support large plankton populations. The plankton are eaten by krill (shrimplike crustaceans) which in turn are eaten by penguins and crabeater seals. The source of DDT for these birds and animals is the ocean water off Antarctica's shore.

How could this pesticide reach such remote parts of the world? DDT has several physical properties that help to assure its distribution: It is persistent in the environment because it is stable—that is, it resists biological degradation; it is volatile and therefore can enter the atmosphere as a vapor; it is more soluble in fats than in water, making its accumulation in the fatty parts of plants and animals possible. When DDT is sprayed onto a crop, it has been estimated that approximately 50 percent of the pesticide enters the atmosphere; the remainder is taken up by plants, finds its way into the soil, or is carried to rivers and streams in surface runoff and adsorbed onto particles of sediment. During periods of heavy rainfall there may be an increased amount of DDT in surface runoff. DDT levels have been monitored in U.S. rivers, but what has been found accounts for only a small proportion (about one percent) of the total amount of DDT sprayed.

The atmosphere, on the other hand, has become a much larger reservoir of DDT, and the pesticide can be readily transported through it. During a storm in January 1965, pesticide-laden dust was picked up by the wind in Texas and dropped the next day in rain over Ohio. Five to six tons of dust per square mile were deposited on Cincinnati; the dust contained DDT, DDE, chlordane, and traces of heptachlor epoxide and dieldrin—all hydrocarbons used as pesticides.

It is not yet known how much of the DDT that vaporizes during and after spraying is broken down in the atmosphere, but the amount is considered to

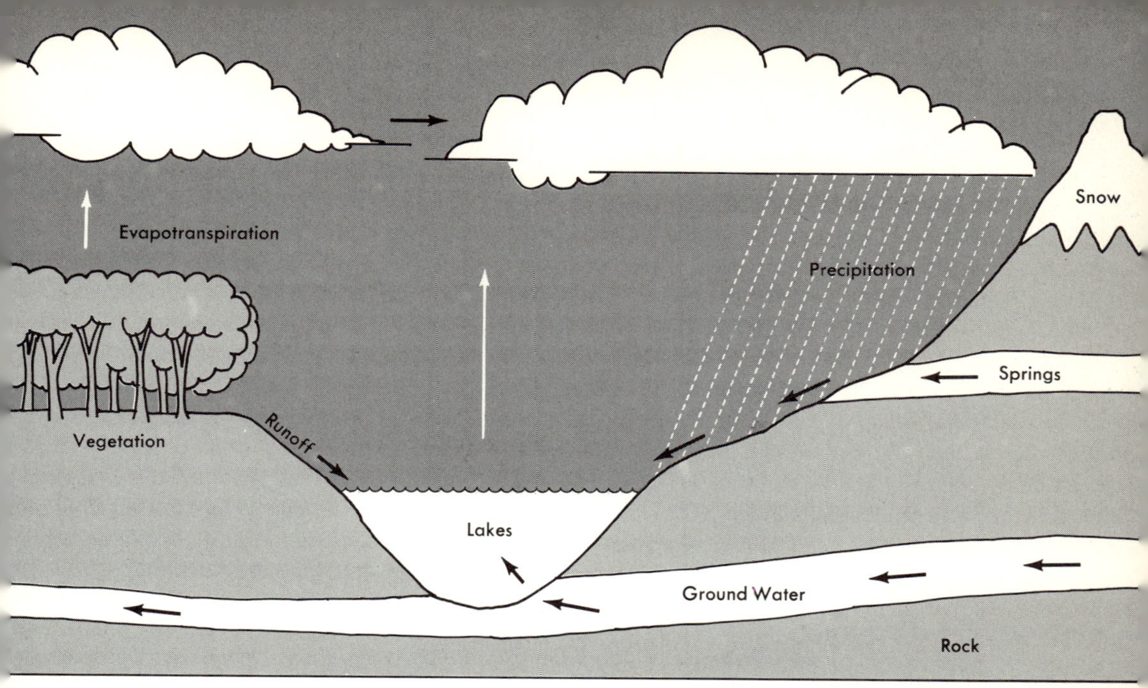

be a small. Rainfall is thought to be the major means by which DDT is removed from the atmosphere and deposited into the oceans. Much of the DDT in the oceans is taken up by plankton. Pesticides not absorbed by the plankton may, after a period of time, settle into the ocean's abyss—where they can remain for hundreds or even thousands of years—and eventually recirculate [5].

It has been estimated that the biota (the living part of the environment) contains approximately 1/30 of the amount of DDT produced in any one year during the 1960s [6]. Even this relatively small proportion of DDT has been significant in disrupting food webs and in reducing some species because the pesticide is circulated around the globe in the atmosphere and passed on to the ocean and taken up from the ocean by many forms of life.

Dependence on Water

All living cells have some contact with water. This is true for single-celled plants and animals (algae and protozoans) that live in an aquatic environment and have a ready source of water, as well as for multicellular aquatic and terrestrial organisms that have, through the course of evolution, internalized their watery environment by developing mechanisms for obtaining and conserving water. Living cells not only inhabit water; they consist mainly of water, which makes up approximately 70 to 80 percent of the average cell's content.

The flow of energy in the ecosystem is also dependent on the availability of water, since it is one of the reactants in a number of important biochemical processes. One of these—photosynthesis—is the important first step in direct-

Life is dependent on water in its many forms.

ing the flow of energy through living systems. In the photosynthetic reaction, the water molecule is split by energy absorbed from sunlight via the chlorophyll of green plants. Part of the split molecule yields oxygen; the rest combines with carbon dioxide to form carbohydrates (which are organic compounds). The organic products of this reaction may then become a source of energy for the animals who eat the plants and, indirectly, for other animals farther up the food chain, including humans, who eat animals nourished on the plants.

Water also plays an important role in transporting many of the materials necessary to the life processes because it has such an exceptional ability to dissolve a variety of substances. In man, food and oxygen are brought to the cells via the bloodstream; chemical messages, in the form of hormones, are also carried in the bloodstream. In higher plants, nutrients are transported from the roots to the leaves in water moving through tubular structures called xylem. In the same way, waste products are removed by means of water; the kidneys remove urea, a waste product of protein metabolism, in urine. To those organisms capable of sexual reproduction, water is the means by which the sperm, mobile only in water, can reach the egg to fertilize it.

Water thus is not only ubiquitous in nature, but is a necessary ingredient for all life. People, no less than protozoa, are dependent upon water. The human embryo is approximately 90 percent water, and adults are 65 to 70 percent water; the percentage of water then decreases in old age. More than one-half of the water in humans is found in the muscles and in the skin; cartilage and bone contain little water. The body very successfully regulates the amount of fluid it contains. For example, water intake depends in part on thirst, which is regulated by the nervous system. Another point of control is at the kidneys: Of the approximately 170 liters of water which pass through the kidneys per day, 168.5 are reabsorbed [7]. The basic biological need for water has strongly affected the history of human civilization. Man has been able to settle only near sources of water which are so essential not only for him but for all the other plant and animal species on which he depends.

Life's dependence on water is the result of millions of years of evolution during which living cells became adapted to such special properties of water as its ability to moderate temperature, to dissolve a great number of substances, and to exist as a gas, a liquid, and a solid at temperatures common to the earth's surface. These properties are all a reflection of the unique molecular structure of water. L. J. Henderson's famous book about the suitability of these properties for the support of life is called, appropriately, *The Fitness of the Environment* [8]. Understanding the molecular structure of water is therefore helpful in understanding the environmental role of water.

The Chemical Structure of Water

While most chemical symbols are unfamiliar to people who have not studied chemistry, "H_2O" has become part of everyday language as a synonym for water. There is more to be understood about the structure of water, however,

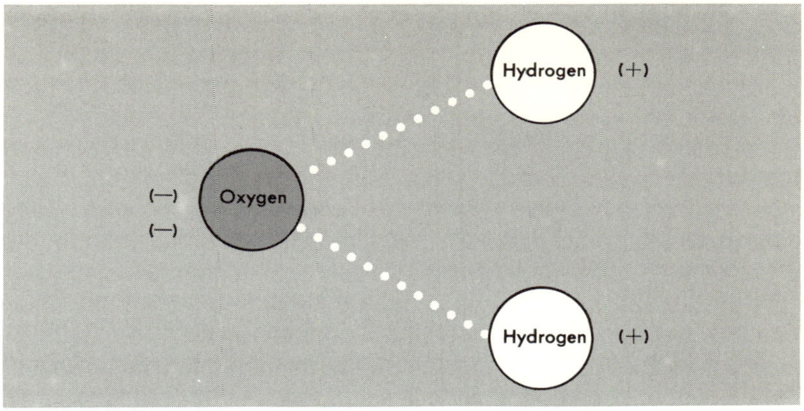

than the simple fact that each water molecule is composed of two hydrogen atoms and one oxygen atom. The nature of the bond that holds these atoms together is important, and even more important is the nature of the bond that is formed *between* molecules of water, for this bond determines water's special properties.

The nucleus of each hydrogen atom contains one proton and is orbited by one electron; the nucleus of each oxygen atom contains eight protons and is orbited by eight electrons. (Protons carry a positive electrical charge and electrons carry a negative charge.) When two hydrogen atoms and an oxygen atom react, forming a water molecule, the atoms become united by sharing a pair of electrons. This is called a *covalent bond.* (The valence, or combining capacity of an atom, is determined by its *outer electron shell*—the paths of the electrons at the greatest distance from the nucleus. This outer shell is sometimes called the *valence* shell. When two atoms share electrons, their valence shells are thus joined by a covalent bond.) In the resulting water molecule, however, the electrons are not shared equally; the oxygen nucleus, with its eight protons is more attractive to the negatively-charged electrons than the hydrogen nucleus, with its single proton. A negative charge therefore accumulates around the oxygen atom in the molecule, and a positive charge accumulates near each of the hydrogen atoms, as shown above. The positively-charged regions (the hydrogens) of one water molecule are attracted to the negatively-charged region (the oxygen atom) of *another* water molecule, and a strong bond, called a *hydrogen bond,* develops between the water molecules. Hydrogen bonds are significant both in their strength and in their dimensions and directions which result in water molecules being joined together in a three-dimensional structure.

Characteristics of Water

Soluble. Water dissolves more substances than any other liquid and, for this reason, is sometimes called the universal solvent. Its molecular structure makes it particularly efficient in dissolving substances with molecules held together by *ionic bonds* (an ionic bond is formed in a chemical reaction when electrons are exchanged rather than shared). When sodium and chlorine react, sodium gives up negative electrons and becomes a positive ion; chlorine gains these electrons and becomes a negative ion. The attraction between the positively-charged sodium ion and the negatively-charged chlorine ion holds the two atoms together in a crystal of sodium chloride. Dissolved in water, the positive and negative ions are attracted to the positive and negative regions of the water molecule. The attraction of the sodium and chlorine ions to each other is overcome to a degree by their attraction to water, effecting their separation.

Water's importance as a solvent is enhanced by the fact that chemically it is a generally *inert* solvent: Its properties are not changed by most of the substances it dissolves, nor does it change most of the substances which dissolve in it. Such a powerful solvent as an acid, on the other hand, will react with a compound which water would simply dissolve. After dissolving sodium chloride, for example, water could regain its former purity (evaporation would separate the water from the salt); one acid, however, could not be freed of its impurities without undergoing another chemical reaction. This is very important when considering biological systems in particular. Because, in general, water is chemically inert, it can transport various essential substances, dissolved but otherwise unchanged, into living cells. Most chemical reactions within plants and animals take place only when substances are in solution.

Renewable. The fact that water remains essentially unchanged also explains why it is a *renewable* resource. A river may carry a heavy load of pollutants in suspended or dissolved form down to the sea, yet when water evaporates from the surface of the ocean under ordinary circumstances it is essentially pure and reusable.

Cohesive and Adhesive. Because of its hydrogen bonding, water also has *cohesion* (the ability to stick to itself) and *adhesion* (the ability to stick to some solid substances)—properties that are important in nature [9, 10]. Except for mercury, water has the highest surface tension of all liquids because of the strong attraction of the polar water molecules on the surface for each other—that is, their strong hydrogen bonds. The formation of a water droplet demonstrates water's cohesion; the droplet would be perfectly spherical except for the force of gravity, and a sort of membranelike film is formed around it by surface tension.

The surface of lakes and ponds provides a habitat for a variety of algae, protozoans, and insects that are able in one way or another to take advantage of water's surface tension. The ability of insects such as water-striders, or pond-skaters, to move on top of water without breaking the surface is one example. Mosquito larvae living underneath the water and hanging upside down from its surface film are another example.

Water adheres to or wets glass because the hydrogen in the water is attracted to the oxygen in the glass without losing its internal cohesion. When water is placed in a glass tube it therefore moves up along the glass sides of the tube, like kerosene up a wick in a kerosene lamp. This phenomenon is called *capillary* or *wick action.* Water moves upward through the soil on the same principle—its cohesion and its adherence to clay particles provide capillary action. Water can move upward as much as several feet in order to be available to the roots of plants.

Thermal Properties of Water

Water has an exceptional ability to store heat. This means that, compared with other substances, a large amount of heat must be applied to water to produce a given rise in temperature. Water's specific heat (the number of calories required to raise one gram of a substance one degree centigrade) is 1.0

How the relationship among three groups of algae may shift as the water temperature rises is shown here. These types of algae may all be found together in water where temperatures range from 68–104°F. In this experiment, as the water temperature rose from 68°, the number of species of diatoms decreased and the number of blue-green species increased. Species of greens also increased with rising temperature up to 86° and then dropped off again. Present evidence suggests that blue-greens are not as suitable as diatoms to the organisms dependent on algae as a food supply. Many blue-greens produce an unpleasant odor and some are toxic to shellfish.

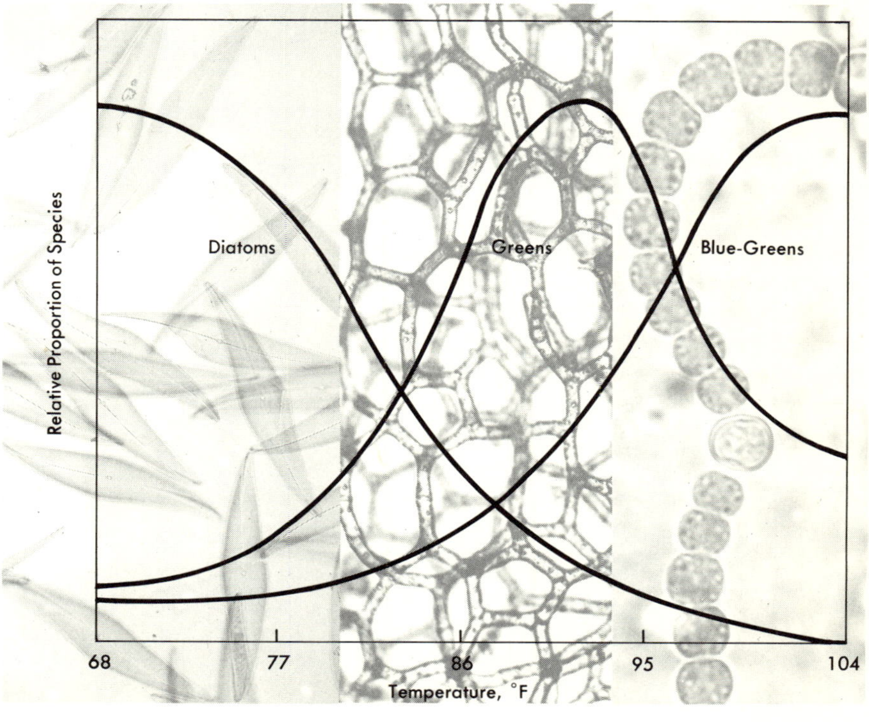

at 15.0° C—higher than the specific heat of any other substance except lithium and ammonia. Below, the specific heat of some other substances are compared with that of water at 0° C [11]:

Water	1.007
Alcohol (ethyl)	0.535
Acetone	0.506
Sulfuric acid	0.270
Carbon tetrachloride	0.198

Because so much heat must be applied in order to raise water temperature, large bodies of water do not fluctuate in temperature as much as large bodies of land do. This is also one reason why shoreline environments have a more constant temperature than inland areas; another reason is the cooling effect of water evaporation. Terrestrial animals have had to adapt to a wider range of temperatures and more rapid temperature fluctuations than aquatic organisms have. Or, to put this another way, aquatic organisms are less adaptable and more vulnerable to temperature changes than terrestrial organisms; because of the protection of their watery environment with its more constant and more slowly-changing temperature, aquatic organisms have not had to develop such adaptability. Water's high specific heat also explains other climatic phenomena—for example, why deserts become cold at night. Since there is little moisture in desert air and sand has a specific heat of 0.2 (one-fifth the capacity of water for storing heat), less heat reserve is available in the desert than in wet climates after the sun sets [12].

One way increased temperatures place stress upon aquatic life is illustrated by the dramatic slope of this line. It shows that water's ability to hold dissolved oxygen decreases as it is heated, making less oxygen available for respiration by aquatic organisms.

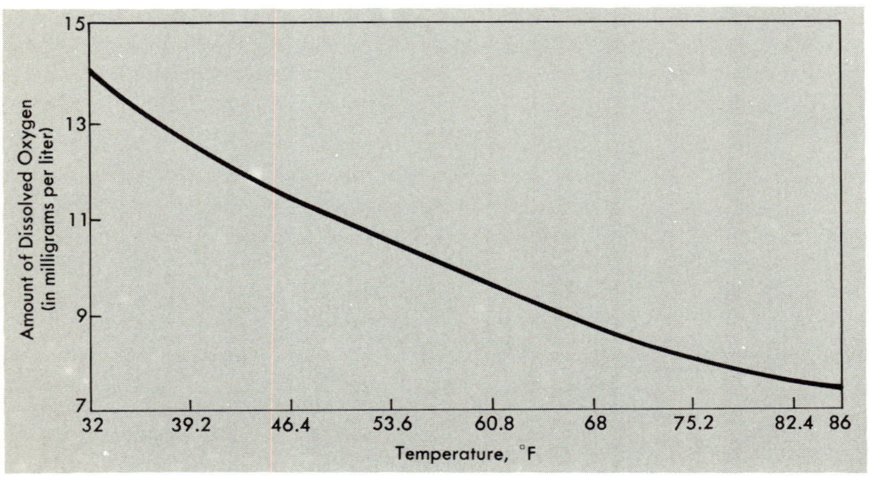

Like other substances, water may exist in three states—solid, liquid, and gaseous—but unlike most other substances, water is found in all three states within the ranges of temperature and pressure common to the earth. A great deal of heat energy is required to evaporate water, changing it from the liquid to the gaseous state. (This latent heat energy is then released when the water vapor returns to the liquid state.) Water is therefore said to have a high *latent heat of vaporization* compared to other substances. Similarly, it takes more heat to melt ice, changing water from its solid to its liquid state. (Again, the liquid water contains latent heat energy which is released if it returns to its solid state.) Water is therefore said to have a high *latent heat of fusion.* The table below [13] compares water with some other substances:

	Heat of Vaporization (calories required to evaporate one gram)	Heat of Fusion (calories required to melt one gram)
Water	539.55	79.71
Alcohol (ethyl)	204.0	24.9
Acetone	124.5	23.4
Sulfuric Acid	122.1	24.0
Carbon Tetrachloride	46.4	4.16

These properties make it possible for water to provide a sort of protective barrier against heat for the organisms that live in it. So much energy is required to evaporate perspiration from the human body, for example, that relatively little evaporation can keep body temperatures from increasing to the lethal range. The energy required for evaporation also helps to moderate the temperature of large bodies of water.

The unusual thermal properties of water are due to its hydrogen bonds. As the temperature of water increases, the thermal agitation of its molecules increases; when the temperature reaches the boiling point, the liquid water is transformed into vapor. Because a great deal of energy is needed to break the hydrogen bonds holding the water molecules together, the boiling point of water is much higher than the boiling points for other closely related compounds. For example, it is 104°C higher than it is for hydrogen telluride, which also has two hydrogen atoms and only a slightly higher molecular weight, but does not form hydrogen bonds. The melting point of water is also comparatively high—51°C higher than that of hydrogen telluride.

How Temperature Affects Density. In nature, solids are usually more dense than liquids and liquids more dense than gases. You would expect ice—the solid phase of water—to be denser than liquid water, but, surprisingly, ice is approximately nine percent less dense than liquid water. As water is cooled, it contracts until it reaches 4°C, and then begins to expand again until it reaches the freezing point, 0°C. As a result, water is at its maximum density at 4°C, and ice or water at any other temperature will float on top of it. This too, is a result of hydrogen bonding: Hydrogen bonds hold water molecules apart; unbonded molecules can pack more closely. In liquid water there are an average of two to three bonds per molecule, while in ice there are four per molecule (that is, all the water molecules are held in the three-dimensional, hydrogen-bonded structure and are therefore less densely packed than in water).

The density properties of water become particularly important when considering the aquatic ecosystem found in a lake. Because cold water is denser than warm water, many lakes (if they are relatively deep) become stratified. The cooler, denser water forms the bottom layer, called the *hypolimnion* (literally, in the Greek from which both "hypo" and "limnion" are borrowed, this means "under the pond," an accurate description). The warmer, lighter, uppermost layer of water is called the *epilimnion* ("over the pond"). The transitional water zone is the *thermocline*, a layer of steeply decreasing temperature. Gases and nutrients do not readily pass from one layer to another. The contents of such a lake are completely mixed during only two periods of the year—spring and fall turnover times. In the spring, as ice melts in temperate regions, the epilimnion becomes even cooler and denser than the hypolimnion; the epilimnion therefore flows to the bottom and vertical mixing occurs. In the fall, the upper layers of the water are again cooled, this time by the drop in atmospheric temperature; the epilimnion becomes denser than the water below it and again mixing occurs.

Heat as a Pollutant

Water, then, is unique in its ability to moderate the effects of heat on the environment. Situations may occur, however, in which so much heat is applied to a body of water that its temperature rises above the level to which the plants and animals in it can adapt. Among the sources of thermal pollution are industrial cooling systems which withdraw water from a stream or lake to cool certain types of industry and electric generating plants and then return the water to its original source at much higher temperatures. Removal of vegetation from river banks removes shade, and the addition of sediment from erosion heats the water because sediment particles absorb light energy. The magnitude of temperature changes that may occur from shade removal and increased sediment erosion was shown in a study carried out in an Oregon logging area. Logging activities along stream banks increased the mean monthly maximum temperatures for July from approximately 57°F in 1965, before logging had occurred, to 71°F in 1967 after logs had been cut and cleared away from the area [14].

However, heat discharged from electric generating plants that can raise stream temperatures 10–30°F is potentially the most dangerous source of

Trees along a river bank provide shade and prevent erosion.

heat pollution. Electric power production has increased rapidly over the years, and it has been predicted that by 1985 a quarter of total surface runoff in the United States will be used for cooling [15].

Plants and animals may be directly affected by heat pollution. For example, as temperature increases, an organism's rate of evaporation increases,

When an area has been logged over, the removal of shade and the increase in sediment from erosion both contribute to an increase in water temperature.

the rate of chemical reaction in the organism changes and important chemicals are destroyed. Effects may also be felt indirectly as a result of changes in the physical properties of water induced by heat pollution. Changes in temperature affect both the density and the oxygen concentration of water, as shown in the following table [16]:

Temperature (°C)	(°F)	Density (grams per cubic centimeter)	Dissolved Oxygen Content (milligrams per liter)
0	32	0.99987	14.6
4	39.2	1.00000	13.1
5	41	0.99999	12.8
10	50	0.99973	11.3
15	59	0.99913	10.2
20	68	0.99823	9.2
25	77	0.99707	8.4
30	86	0.99567	7.6
35	95	0.99406	7.1
40	104	0.99224	6.6

Many biological events—for example, the spring and fall blooms of algae which occur in many lakes—are closely linked to the timing of turnover, or mixing, in thermally-stratified lakes. The addition of large amounts of heated water—thus changing the water's density—could prevent turnover from occurring, particularly in small lakes. Even if turnovers were not prevented, a delay in turnover time in the fall could lengthen the growing season and produce overgrowths of algae; since exchange of gases does not occur until turnover time, and oxygen in the bottom layer becomes depleted as the summer progresses, a delay in turnover time could also have drastic effects on the fish population. Cornell scientists, studying the site of a proposed nuclear power plant in Cayuga Lake in New York, concluded that since the power plant operation would withdraw approximately ten percent of the average volume of the lake and return this water at a higher temperature, the possible changes in stratification pattern could seriously threaten the lake's ecological balance [17].

One of the most important determinants of water quality is oxygen concentration. As temperature increases, oxygen content decreases. The relationship between temperature and dissolved oxygen in natural waters is not this straightforward, however, because a number of oxygen-consuming and oxygen-producing events occur simultaneously. For example, photosynthesis by green plants and surface mixing both add oxygen to the water while bacteria and animals consume oxygen. Oxygen content determines what types of organisms can survive in a given situation. In many cases decreased oxygen, rather than elevated temperature per se, may be the immediate cause of a fish kill or other environmental disasters.

Temperature also directly affects living things. The basis of all life functions is a complicated set of biochemical reactions that are influenced by physical factors such as temperature. In the face of changing temperatures, plants and animals must find ways to maintain fairly constant internal body temperatures, or they must be able to modify their metabolisms so that changes in their internal temperatures will not be harmful. Warmblooded animals—the birds and mammals—can regulate their body temperatures within certain narrow limits when temperature changes occur in the environment. On the other hand, the internal temperatures of coldblooded animals—fishes and reptiles—correspond to external temperature changes.

Extensive research has been done on the effects of temperature on fish metabolisms. The results of these studies are reflected in temperature standards recommended by the Federal Water Pollution Control Administration [18]. For warm-water species, it is suggested that the temperature of a stream not be raised by more than 5°F and that the temperature of the epilimnion of lakes and reservoirs should not be raised more than 3°F. These figures are based on the monthly average of the maximum daily temperature. In addition, normal daily and seasonal temperature variations should be maintained. For cold-water species, no heated effluents should be discharged into inland trout streams, headwaters of salmon streams, or trout and salmon lakes or reservoirs, and the hypolimnion should not be warmed. For other streams, temperatures should not be raised above 5°; for lakes and reservoirs, 3°. The table* below shows the maximum recommended temperatures for various fish species and their associated biota. These standards include suggested temperatures which should not be exceeded in order to maintain growth and reproduction, as well as a healthy adult fish population. The recommended temperatures for species not listed in the table may be established if and when necessary information becomes available.

	Species and Biota
93°F:	Growth of catfish, gar, white or yellow bass, spotted bass, buffalo, carpsucker, threadfin shad, and gizzard shad.
90°F:	Growth of largemouth bass, drum, bluegill, and crappie.
84°F:	Growth of pike, perch, walleye, smallmouth bass, and sauger.
80°F:	Spawning and egg development of catfish, buffalo, threadfin shad, and gizzard shad.
75°F:	Spawning and egg development of largemouth bass, white or yellow bass, and spotted bass.

*From *Water Quality Criteria*, report of the National Technical Advisory Committee to the Secretary of the Interior, Federal Water Pollution Control Administration, Washington, D.C., April 1, 1968, p. 43, Table III–1.

68°F: Growth or migration routes of salmonoids and egg development of perch and smallmouth bass.

55°F: Spawning and egg development of salmon and trout (other than lake trout).

48°F: Spawning and egg development of lake trout, walleye, northern pike, sauger, and Atlantic salmon.

These temperature standards are based on observations and studies showing that fish are governed by upper and lower lethal metabolic limits which are species-specific and that many metabolic changes occur when temperature is raised or lowered. As temperature is increased, both oxygen consumption and heart rate increase. More oxygen is needed at higher temperatures because of a higher metabolic rate and the hemoglobin's lowered affinity for oxygen. At the same time, however, the amount of oxygen dissolved in the water is decreasing. For example, at 33°F carp can survive at oxygen levels as low as 0.5 mg/l. At 95°F the water must contain 1.5 mg/l if the carp are to survive. Swimming speed (which is used as a laboratory test of metabolic rate) increases with increasing temperature. For example, salmon swim twice as fast at 60°F as they do at 35°. As with other biological phenomena, activity may increase with temperature up to a certain point and then begin to decline. The swimming speed of a trout increases until the temperature reaches the range of 49–66°F; the trout then begins to slow down. However, the swimming speed of the minnows on which the trout depends for food is not affected in the same way; when the temperature reaches 70°F, the trout has slowed down to the point that it can no longer catch the minnows [19]. A small temperature difference—only 4°—can therefore spell life or death for the trout, even though the higher temperature in itself would not be lethal.

Disease resistance is also linked to temperature. Not only does resistance to disease decrease with increasing temperature, but increases in temperature also increase the rate of microbial activity. Certain bacteria are particularly lethal to fish. An example is the myxobacterium *Chrondroccus columnaris*. This bacterium is relatively innocuous below 60°F; between 60° and 70°F, it may invade wounds; above 70°F, the bacteria can invade healthy, undamaged tissue. This bacterium was thought to be responsible for a large fish kill of blueback salmon in the Columbia River in 1941. Water temperatures there were exceptionally high during that summer, and on August 20, the river temperature reached an all-time high of 74.5°. Of the 25,000 salmon expected on the migration route, only 949 arrived [20].

Temperature increases may become a barrier to fish migration and, in this way, seriously effect the reproduction of a species. The effects of dam construction in the Columbia River area in the State of Washington has been extensively studied. One study showed that an increase in temperature to more than 70°F delayed passage of salmon from the Columbia River into the

Okanogan River on the way to their spawning grounds. The migration continued when temperatures fell below 70° [21].

When given a choice, fish attempt to avoid lethal temperatures by moving to water of a preferred temperature. For this reason, in winter when they would ordinarily seek warmer water by moving South, fish are often attracted to the warm areas of effluents discharged from power plants. When the power plants are shut down periodically for cleaning purposes, the warm effluent is suddely cut off, there is a drastic drop in local water temperature, and the fish may be killed. In January 1972, the Sports Fishery Institute Bulletin reported the death of 200,000 baby menhadden after the operational shutdown of a power plant in New Jersey caused the water temperature to drop 11°F over a 24-hour period [22]. Fish do have varying abilities to adapt to changes in temperature since they must normally cope with seasonal temperature changes. However, these changes are usually gradual; very rapid changes in temperature can prove fatal.

The survival of any species depends on its ability to reproduce. In some cases, temperature triggers reproductive activity—for example, a rise in temperature may trigger the development of the gonads which, in turn, may signal the deposit of eggs. Yet above certain temperatures, some fish do not reproduce at all. This has been found to be true of the carp: Cell division in carp eggs is prevented at temperatures in the range of 68–75°F [23].

Cataloging the effects of heat (or of any other pollutant) on particular species of plants and animals does not give a complete picture of the impact of that pollutant on the environment. While a decrease in one species—or even its complete elimination—may not seem drastic, each species occupies an important place in an ecosystem and the effects of its removal will be felt in other parts of the system. A change in predator-prey relationships—such as the inability of the trout to feed on its natural prey, the minnows—is a case in point. When a predator dies out, the population of its prey increases, producing further effects on the system's food chains and energy relationships.

Because of these complexities, there have been attempts to assess the ecological effects of power plants—operational or proposed—rather than simply the effects of direct heat on a list of aquatic organisms. One such study was made some years after a power plant was built 15 miles from the mouth of the Connecticut River. A number of biological aspects of the river were examined over a five-mile stretch downstream from the plant. Microbiological investigation showed that diatoms had been replaced by blue-green algae, while the investigation of bottom life showed a decline in the diversity of species as well as in the number of individuals per species. The decrease in species number was thought to be due to the increased velocity of water at the plant intake. Deposition of silt at the discharge point had caused a decrease in the clam population. Extensive observations of the fish revealed that neither the resident nor the migrating fish populations seemed affected. While the heated plume of the power plant effluent extended to the opposite bank of the river from the point of discharge, juvenile shad moving downstream to the sea passed under the heated water. The investigators concluded that the power plant had produced no major short-term detrimental effects, but that its long-term ecological effects could not be predicted [24].

Water Is Not Pure in Nature

Approximately half of the known chemical elements have been found dissolved in natural water [25]. A clean running stream, even in its unpolluted state, contains a complex mixture of organic and inorganic substances. In fact, living things could not survive in pure, distilled water.

Salinity. Although water is essentially pure when it is evaporated, many substances become associated with it as it travels through the atmosphere, and when water returns to the earth as rain it contains a number of salts, most of them in trace amounts. As these salts are dissolved in water they separate into electrically-charged particles called *ions*. These ions include chlorides, bromides, iodides, sulfates, phosphate, bicarbonate, sodium, calcium, magnesium, potassium, and nitrates [26]. Salts and minerals are dissolved in water when precipitation reaches the earth's surface and water flows over rocks and through the soil. Every body of water flows over rocks and through the soil. Every body of water has a somewhat different chemical composition since each comes into contact with different geological formations.

Fresh water is "hard" when it contains ions which form an insoluble curd with soap, making the soap ineffective as a cleanser. Calcium and magnesium are the main components responsible for the hardness of water. Aluminum, barium, zinc, strontium, iron, manganese, and other ions that may cause hardness are usually found in small quantities under natural conditions.

Bicarbonate is the most abundant ion found in U.S. streams, mainly because of the solubility of limestone (calcium carbonate) and the action on limestone of the carbonic acid in water, which we will discuss later. Sulfates, calcium, sodium, and chlorides are found in lesser amounts. In general, fresh water is about one-tenth of one percent salt solution, while sea water is 3.5 percent salts by weight, with sodium and chloride ions most abundant. Calcium carbonate in the oceans has been incorporated into animal structures, such as the shells of marine animals, and is found in sediment rather than dissolved in the water.

Since plants and animals have become adapted through the course of evolution to the mixture of substances found in their environment, a change in the concentrations of these substances in water can have serious effects. While many chemicals are necessary for normal cell function, certain chemicals are toxic in large quantities. Moreover, an increase or decrease in *total salinity*, regardless of the kind of salt present, can also upset cellular function. Thus, freshwater animals have developed mechanisms for living in external environments that contain fewer salts than their internal environments; marine organisms, on the other hand, have adapted to life in environments that contain higher concentrations of salts than their bodily fluids do. Organisms that can tolerate the greatest degree of change in salinity are those that inhabit estuaries—a special type of aquatic environment at the mouths of rivers that is subject to constant fluctuations between fresh water and sea water.

Taking into account these limitations to biological adaptability, especially the ability to regulate internal salt concentrations, the Federal Water Pollution Control Administration of the U.S. Department of the Interior recommended standards [27]: For estuaries, no changes should be made in channels, basin geometry, or freshwater inflow that will create more than a 10 percent change from natural variation in salinity. In order to maintain a diversified freshwater animal population, the salt concentration should not exceed 1500 mg/l. Also, total dissolved solids should not increase to more than one-third of the water's natural concentration in order to maintain local conditions. These standards are important because salts from a number of human activities find their way into surface and ground water. Some of the sources of excess salinity are effluents from desalination plants, brine from deep-well injections, and salt from the winter deicing of streets in urban areas.

While the desalination of salt water for drinking purposes is not a common practice, it is now carried out in approximately 25 communities around the world. The process involves separating salt from water and results in two streams of water—one which is less saline than the original and a waste stream which is more saline than the original. An additional waste stream contains heated cooling water. Some plants have been planned which will have conversion capacities of 50 million gallons per day. A study of a government-sponsored saline conversion plant at Point Loma, near San Diego in California, pointed out some of the deleterious effects this type of effluent can have on the environment. The concentration of the brine discharged was 1.5 to two times that of sea water. The temperature of the cooling water effluent was also high. Sea water temperatures varied from 52°F (winter) to 72°F (summer); water temperatures in the effluent stream ranged from 83°F (winter) to 110°F (summer).

Examination of aquatic life showed a significant drop in the numbers of plant and animal species, particularly at the conversion plant outfall. The mean number of species in two sample areas near the plant outfall was 4.2 and 5.0; in three adjacent areas more distant from the source of effluent, the average number of species was 7.7, 8.0, and 12.0. Thus, only the more hardy species could survive the combined effects of increased salinity and temperature [28].

Since 1947, the use of salts for the winter deicing of roads and streets has increased 1800 percent, from a half-million tons in 1947 to nine million tons in 1970. A study of the salt concentration in effluent from the Kennedy Expressway in Chicago by a member of the American Public Works Association revealed that, assuming 1.5 tons of salt are applied for each mile of a two-lane highway, there would be 1300 mg/l of chloride ion in the runoff. Thus, the chloride ion alone in highway runoff may be close to the 1500 mg/l salinity limit recommended in the previously cited FWPCA standards for fresh water [29].

While most of the major U.S. rivers recently analyzed for chloride concentrations have shown little increase over previous concentrations, ponds and lakes are another story. For example, in a study of Maine roadside farm ponds from 1966 to 1968, average concentrations as high as 110 mg/l of chloride were found, where the normal concentration had been 5–10 mg/l [30].

Irondequoit Creek and Irondequoit Bay in Rochester, New York, have shown a tenfold increase in chloride concentration since 1910. This increase can be correlated with changes in the stratification pattern of the bay. In the spring of 1970, complete mixing of the bay did not occur (that is, mixing did not reach to the very bottom) because of increased density due to salinity. The fall turnover occurred approximately one month late. During the winter, at a depth of 21 meters, chloride concentration was as high as 400 mg/l and bottom water was without oxygen [31].

Salt contamination also poses a threat to ground water supplies. According to 1960 figures, just a little less than one-fifth of the total water used in this country was ground water [32]. Yet each day an estimated ten million barrels of oil-field brine are injected underground, and about 200 deep wells are used for industrial waste disposal [33]. The great danger, of course, is that, while engineers claim that these liquid wastes are being stored within safe geological formations (the area where injection occurs should be completely surrounded by impermeable rock), no one knows for sure where these wastes ultimately will travel. The dangers are many [34]: Even seemingly impermeable rocks could have unpredicted flaws through which liquids might seep under pressure; natural disasters, such as earthquakes, could release millions of gallons of wastes; corrosion or damage of a well shaft would allow liquids to be forced back up to the surface. And these are only a few of the possibilities. Ground water moves much more slowly than surface water: Ground flow is usually measured in feet per year, while river flow is measured in feet per second, and in an underground aquifer the turnover rate for wastes would be 200 to 10,000 years [35]. Thus, a large portion of the water supply available for future generations could be contaminated.

Turbidity and Sedimentation. Not all of the impurities found in water are dissolved substances. Water's action as an erosive agent accounts for the large assortment of suspended materials it contains. Suspended particles, visible to the naked eye or under a microscope, are responsible for the turbidity or cloudiness of water; unlike dissolved substances, suspended materials eventually settle to the bottom of a lake or stream. Sand, silt, clay, organics, algae, and bacteria are some of the substances that contribute to turbidity. Surface runoff is the earth's most important erosive force. Again, water's special physical and chemical properties make this possible: Rocks are split when the water which fills their cracks expands upon freezing; because of its high surface tension, a drop of rain is an effective shattering agent; because of its bouyancy, water can carry eroded materials many miles from their source.

The largest portion of suspended material found in U.S. waters is the re-

Most of the suspended material found in the nation's waters is the result of soil erosion, and much soil erosion is created by land clearing for urban and suburban developments and highways.

sult of soil erosion. Roadsides produce 56 million tons of sediment a year, and three billion tons of soil are displaced from cultivated land and overgrazed granges. Erosion from the construction of suburban housing developments is a relatively new problem, but may contribute more sediment to the water than farmland erosion. For example, for each square mile of construction activity in a suburban development area in Fairfax County, Virginia, 25,000 tons of sediment were deposited in a downstream lake [36].

Heavy sedimentation has far-reaching effects on plant and animal life. Plants need sunlight to carry on photosynthesis and can grow only to depths where light intensity is great enough. Because water is transparent, light may be available for photosynthesis at depths of approximately 5–90 feet; in Lake Tahoe, a definite light penetration has been measured at a depth of 136 feet. However, heavy turbidity, by scattering and absorbing light, disturbs the light-transmitting quality of water and reduces plant production [37].

Sedimentation effects water quality in a number of other ways. By absorbing light energy, silt particles may increase water temperature. When silt and other solids settle to the bottom of a stream, they can smother bottom life. In one experiment, freshwater mussels were smothered when one quarter to one inch of silt was allowed to settle over either a sand or a gravel bottom habitat to which they had become adapted [38]. Silt sometimes causes algae to flocculate (that is, to make them stick together) so that they no longer float. When the algae sink to the stream bottom, they increase the amount of oxygen available to the increased number of bacteria that degrade this material [39].

The same qualities that make water effective as an erosive agent make it useful to people as a cleaning agent. A group from Marquette University reported to the 1970 federal hearings on phosphates and eutrophication on the difference in cleaning power between a non-phosphate detergent (Heinecke's B-5), a phosphate detergent (Tide), and plain water. They took diapers soiled with garden dirt and barbecue sauce and dried them, and then placed them in washing machines. Plain water cleaned better than either of the two detergents. Even taking into account that no tests for grease (soap or detergents are essential to remove grease) or bacteria were made, the experiment illustrated water's important role in cleaning—a process that is usually attributed to the cleaning agent alone [40].

Gases. Gases that become dissolved in water originate from two main sources: the atmosphere and the byproducts of the metabolisms of living organisms. The air contains 78 percent nitrogen, 21 percent oxygen, and 0.03 percent carbon dioxide. The amount of each of these gases that will be found in water depends on the relative amount of that gas in the atmosphere, whether or not the gas reacts with water, and its *solubility coefficient* (the liters of gas absorbed by one liter of water at 0°C and one atmosphere pressure). Oxygen and nitrogen have comparatively low solubility coefficients (0.049 for oxygen and 0.024 for nitrogen) and are not found in large amounts in the water. The solubility coefficient of carbon dioxide, on the other hand, is 1.713. In addition, carbon dioxide reacts with water. Thus, compared to the small amount of carbon dioxide found in the air, a relatively large proportion is found in the water. Biological processes contribute gases to the water also. Oxygen is a product of photosynthesis; carbon dioxide is a product of respiration; and hydrogen, methane, ammonia, and hydrogen sulfide are all products of the bacterial decomposition of organic matter.

pH and Carbon Dioxide. The pH of any body of water describes its degree of acidity or alkalinity. The pH scale, which ranges from 0 to 14, measures the concentration of hydrogen ions (hydrogen atoms with a positive electric charge) and hydroxyl ions (an oxygen and a hydrogen atom with a negative charge). If a solution has equal numbers of hydrogen and hydroxyl ions, it is neutral—midpoint on the pH scale, or 7. More hydroxyl ions increases the alkalinity. (The scale is logarithmic, so a pH of 5 is ten times more acid than a pH of 6; a pH of 9 is ten times more alkaline than a pH of 8.) Pure, distilled water has a pH of 7, and the pH of most natural waters is 7 to 9 (neutral to slightly alkaline). Water at pH7 consists mainly of undissociated water molecules (H_20) and a very small number of molecules that have separated into hydrogen and hydroxyl ions.

Natural waters are *buffered*—that is, there is a natural mechanism which regulates the pH so that it usually varies within the narrow range between 7 and 9. The concept of buffers also applies to the regulation of pH in the internal environment of living things. The pH of human blood, for example, is maintained at 7.4.

The carbon dioxide and the bicarbonate ions dissolved in water are mainly responsible for the regulation of pH in lakes and streams. If a strong acid is added to distilled water, which is without these dissolved compounds and is not buffered, its pH will change drastically; if the same amount of acid is added to naturally buffered lake water, however, the change in pH will be relatively small.

Rain picks up carbon dioxide in the atmosphere. The water and the carbon dioxide react to form carbonic acid, and carbonic acid becomes dissociated into hydrogen and bicarbonate ions. This increase in hydrogen ion concentration gives rain a pH of 5.7 under normal conditions. If that were the end of the matter, we would expect lake and stream water to be similarly slightly acidic. But after water reaches the ground it percolates through the soil, acquiring more carbon dioxide which has been produced by the respiratory activity of plant roots and microorganisms. When water containing carbon dioxide (and carbonic acid by the association) comes into contact with rocks such as limestone, calcium bicarbonate is leached out and set free in the water.

As mentioned previously, the bicarbonate ion is the most abundant ion found in natural fresh waters. These bicarbonate ions react with the hydrogen ions to form carbonic acid (the reverse of the reaction in which carbonic acid is dissociated into hydrogen ions and bicarbonate ions). Thus, the conversion of bicarbonate ions to carbonic acid (a relatively weak acid) uses up the excess hydrogen ions, thereby buffering the effect of the latter on the water's pH. A strong acid will add many hydrogen ions to the water, and how much acid can be added to any stream or other body of water before the pH is changed therefore appreciably depends on the amount of bicarbonate ions available in the water to take up the added hydrogen ions.

Changes which have recently been recorded in the acidity of rain demonstrate both how changes in pH can occur and how industrial pollutants can affect water on a global scale. In a number of places in or near highly industrialized areas, the pH of rainwater has dropped. Measurements taken in New England (Hubbard Brook Experimental Forest) and New York (Finger Lakes region), as well as in the Scandinavian countries, show pH values as low as 3.0. This increase in acidity is due presumably to sulfuric and nitric acids in the atmosphere, and the source of these acids is the sulfur and nitrogen oxides emitted to the atmosphere by the burning of fossil fuels. In Sweden, changes in the pH of some lakes and rivers has been attributed to acid rain. One possible ecological effect of acid precipitation is an increase in the amount of

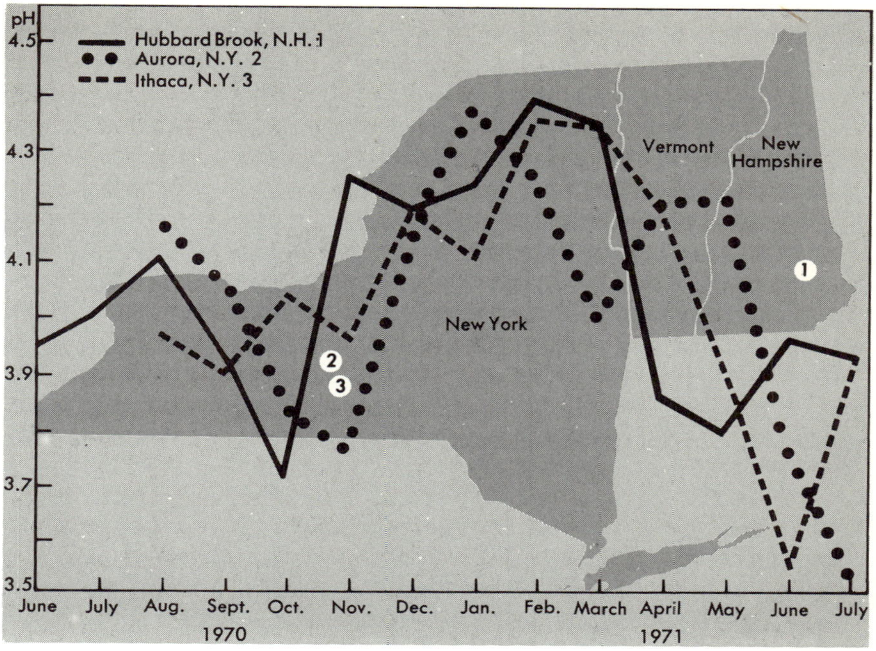

The acidity of rain in the Finger Lakes Region of New York State and at the Hubbard Brook Experimental Forest in New Hampshire is illustrated by low pH values during 1970–1971 at three sites. Annual averages, weighted to take into account the amount of precipitation that fell, were as follows: Hubbard Brook (June to June): 4.03; Aurora (August to August): 3.91; Ithaca (August to August): 3.98. Values of pH below 5 indicate unusually strong acids in rain. Low values have been found in rain in Massachusetts and Connecticut, as well as in New York and New Hampshire.

calcium and nutrients which may be leached from the soil; another is the death of fish; a third is a reduction in forest growth [41].

A more localized source of acid pollution is the drainage from both operable and inoperable coal mines. This changes the quality of water in a number of ways and has seriously affected aquatic life in mine areas. Acid from coal mine drainage is primarily due to the iron disulfide or pyrite which is found in association with coal deposits and which becomes exposed to air and water during mining. When pyrite is exposed to water containing an adequate amount of oxygen, a series of chemical reactions take place which result in the formation of iron hydroxide and sulfuric acid.

Acid streams are either copper-colored or brown due to the iron hydroxide which precipitates (becomes insoluble). The iron precipitates often coat stream beds and smother bottom life.

It has been estimated that 10,500 miles of streams in the Appalachian area of the eastern United States have been affected by mine drainage; approximately half of these have been rendered acidic. Underground mines contrib-

Almost as lifeless as the mountains of the moon are these mountains of the mine, and acid drainage into the water makes life difficult or impossible for aquatic organisms.

ute 58 percent of the drainage, surface mines, 28 percent, and 14 percent comes from a combination of the two [42].

The story of Beaver Creek Basin in south central Kentucky illustrates some of the changes in both the chemistry and the biology of streams which occur as a result of mining. Mining activity took place in parts of the Basin in

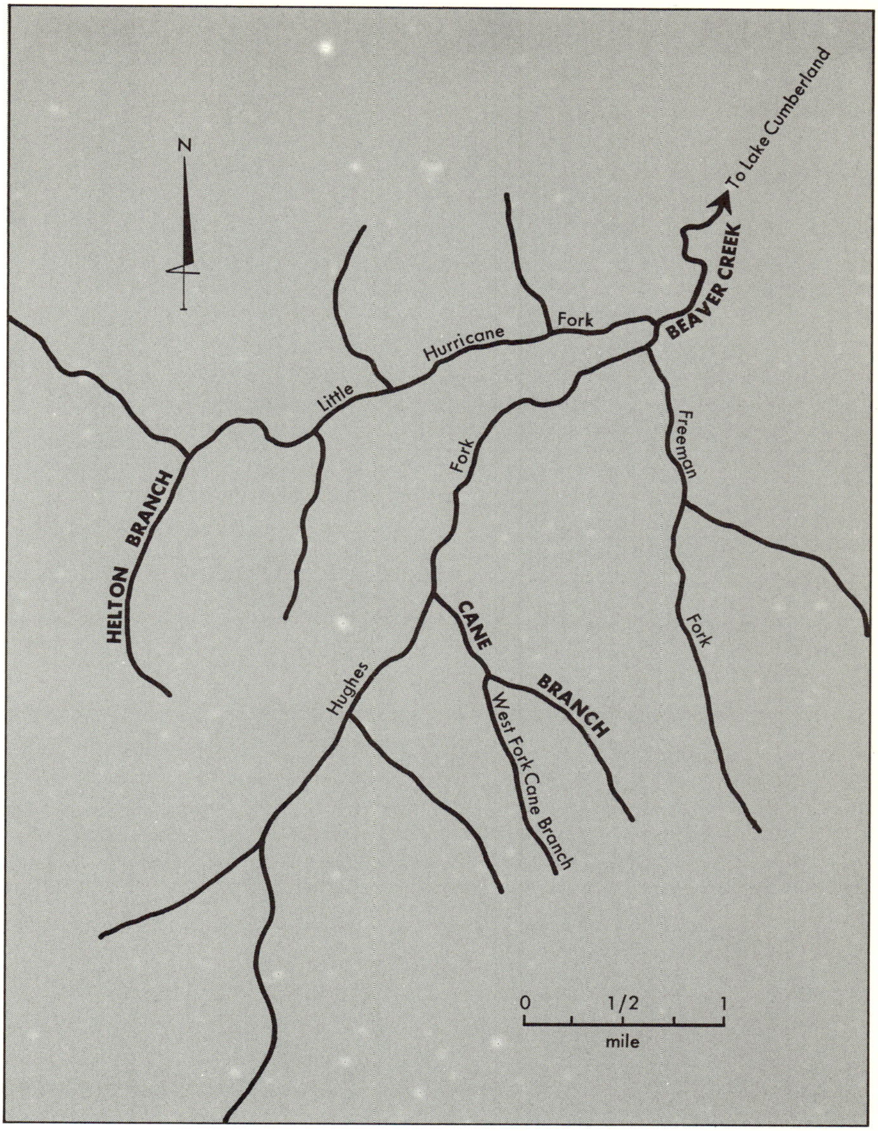

Beaver Creek Basin, showing Helton Branch, which received no mine drainage, and Cane Branch, which was polluted from mine operations. Mine drainage altered the chemistry and biology of Cane Branch.

1955, 1956, and 1959, and a study of the area was carried out by the U.S. Geological Survey between 1955 and 1966 [43]. The Cane Branch of the Hughes Fork (see the map above), which had received mine drainage, and the Helton Branch of the Little Hurricane Fork, which had not, were compared. In the Helton Branch, the dissolved solids content ranged from 15 to 20 mg/l and the pH was 5.8 to 7.5. The dissolved solids consisted mainly of calcium and bicarbonates—50 percent—as well as 25 percent silica, 15 percent sulfates, and a ten-percent mixture of other ions. In the Cane Branch, the dissolved solids concentration ranged from approximately 30 to 400 mg/l and the pH was 2.5 to 4.35. Between 1957 and 1962, Cane Branch transported 1367 tons of dissolved solids per square mile of drainage area; the Helton Branch transported 111 tons per square mile. *No* bicarbonate ions (so important to buffering capacity) were found in the Cane Branch [44]. The following tables show some of the differences in chemistry between the two branches:

Net Dissolved Solids Load*

Year	Tons (per square mile per year) Cane Branch (with mine drainage)	Helton Branch (without mine drainage)
1957	194	13
1958	189	26
1959	165	3
1960	382	24
1961	196	19
1962	241	26

Chemical Properties of Streams in Beaver Creek Basin, August 4–5, 1964

	Cane Branch	Helton Branch	Upper Beaver Creek Basin, Station 12
pH	3.2	6.6	>6.0
Specific Conductance (in μmhos)†	689	19	28
Hardness (calcium) (mg/1)	178	9	11
Sulfate (mg/1)	242	.4	2.4
Bicarbonate (mg/1)	—	10.0	12

The two tables above are adapted from Tables 6 and 7 in J. J. Musser and R. J. Pikering, "Geochemistry of Water," in *Influences of Strip Mining on the Hydrologic Environment of Parts of Beaver Creek Basin, Kentucky, 1955–1966*, Geological Survey Professional Paper (Washington, D.C.: U.S. Government Printing Office, 1970), p. C38.

*Dissolved solids transported minus dissolved solids in precipitation.

†A measure of the electrical conductivity of water, which increases with increased salinity.

The increased mineral content in the Cane Branch reflects the accelerated weathering process which results from high acidity.

As shown on the preceding maps, both the Helton and Cane branches meet to form Beaver Creek three miles downstream. (Data for Beaver Creek also appears in the table opposite.) Chemically and biologically, Beaver Creek approached the control situation—pH was normal there because the stream's acidity had been diluted and neutralized by the bicarbonate alkalinity of incoming tributaries. Sulfate and hardness were only slightly higher than the contributing streams.

Increased acidity and mineral content also affected wildlife. A number of biological parameters were investigated. Bottom fauna (animal life) in the two branches of the creek were compared: the Helton Branch contained an average of 178 bottom organisms per square foot; the Cane Branch contained a mean of 30 organisms per square foot. The variety of insect orders changed also, as shown below [45]:

Year	*Cane Branch*	*Helton Branch*
1959	6	7
1960	3	8
1961	2	7
1962	2	7
1964*	2	7
1965	2	8

May flies and caddis flies, good food organisms for fish, disappeared almost entirely from the Cane Branch while these two species made up 28 percent of the total insect samples collected in the Helton Branch. In August 1964, the abundance of eight species of fish was measured (in pounds per acre) at six stations—three on streams affected by mine drainage and three on unaffected streams [46]. At one of the stations on an affected stream, 8.9 pounds per acre of creek chub were found, as were traces of arrow darter and striped darter; otherwise, there were no fish at the stream stations. In contrast, at the stations on unaffected streams there were 15.9 pounds per acre of creek chub and arrow darter at Helton Branch, and 33.6 pounds per acre of creek chub, hogsucker, white sucker, and southern redbelly dace and a trace of rock bass at Little Hurricane Fork.

In addition to a general reduction in the variety and quantity of organisms, an important biological indicator of acid conditions is the presence of acid bacteria [47]. *Ferrobacillus ferroxidans,* which grows best at pH 3.5 and utilizes ferrous sulfate, was isolated from Cane Branch during all four seasons of the year from 1966 to 1968; from the Helton Branch, it was isolated during only one season per year—the summer. This bacteria produces ferric hydroxide and sulfuric acid, further adding to the acidity of the water.

*Data unavailable for 1963.

Synergism and Antagonism

To further complicate the important role physical and chemical characteristics play in sustaining life in the aquatic environment, we must add the problems which result from interactions between the various factors considered thus far. In some cases, the combination of two or more chemicals results in an *additive* effect; that is, the effect can be predicted by simply adding the toxicities of the combining chemicals. In other cases, interactions between the chemicals, or often between one of the chemicals and the target organism, make the combination more toxic than can be predicted and result in a *synergistic* effect. In a third kind of interaction—called an *antagonistic* effect—the toxicity of one of the chemicals is reduced. Appendix Table 1 summarizes the ways in which various factors influence the toxicities of a number of chemicals introduced into water by industry. This summary of experimental findings is not intended to be a complete description of water chemistry, but only an aid in understanding some of the ways in which physical, chemical, and biological factors interact to affect toxicity. It is worth looking at two of the heavy metals in a little more detail here.

Heavy metals are sometimes added to water by leaching from mineral-bearing rock and soil, but increasingly in recent years they have reached water from industrial emission. There has been much research designed to determine the toxic effects of these metals on aquatic organisms; as we have seen, this toxicity may be modified by a variety of factors: the water's hardness or temperature, the presence of other chemicals, or a combination of one or more of these.

Zinc is a good example. One of the major determinants of the toxicity of zinc and other heavy metals is the water's degree of hardness; as hardness (the amount of calcium carbonate) increases, toxicity usually decreases. In an experiment testing the toxicity of zinc to rainbow trout, fish exposed to water with 5 ppm zinc and 320 ppm calcium carbonate survived about three times as long as fish exposed to same concentration of zinc but only 50 ppm calcium carbonate [48]. When zinc appears in water in combination with the salts of other heavy metals, however, additive or synergistic effects may result. For example, the toxic properties of zinc and cadmium are additive, while the properties of zinc and nickel and of zinc and cyanide are synergistic. Both copper and zinc are highly toxic, and together, especially in soft water, they become far more toxic. Fish that survived in water with 8 ppm copper alone or in water with 0.2 ppm copper alone, were killed by water with a combination of one ppm zinc plus 0.025 ppm [49].

Biological as well as chemical interactions in the environment can be responsible for increasing the toxicity of a given substance. Mercury is a case in point. It is found in a variety of forms—quicksilver (elemental, liquid mercury), elemental mercury vapor, mercuric salts (such as mercuric chloride), and organic mercury (mercury which has combined with a carbon compound). Not all of these forms of mercury are equally toxic to people, however. Methyl mercury, an organic form, is by far the most toxic, and inorganic mercury can be transformed into methyl mercury in nature.

Mercury toxicity depends on both the quantity absorbed by an organism and the site of accumulation. Inorganic mercury accumulates in the human body mainly in the kidney, but also in the brain, the liver, and the intestine. Elemental mercury is relatively nontoxic when ingested. However, the inhalation of mercury vapor is extremely hazardous, and, since the vaporization of mercury occurs quite readily, it has been the most serious industrial mercury hazard.

Methyl mercury is much more readily absorbed by the body when it is ingested than is inorganic mercury: 90 to 95 percent of methyl mercury may be absorbed by the intestine, in contrast to only two percent of inorganic mercury. Methyl mercury diffuses through biological membranes much more readily, is more evenly distributed throughout the body, and is excreted much more slowly than other mercury compounds. Brain damage is more likely to be caused by methyl mercury than by inorganic mercury poisoning [50].

The Minamata disaster in Japan brought widespread attention to mercury's potential as an environmental polluter. Here 110 people became the victims of mercury poisoning between 1953 and 1960 as a result of eating fish taken from Minamata Bay where a vinyl chloride plant released methyl mercury as one of its waste products [51]. Since that time, it has been learned that methyl mercury is the most common form of mercury found in the aquatic environment. Fish excrete it very slowly; accumulations of mercury as high as 5000 times the amount found in water have been reported in fish. A study of Lake St. Clair showed that mercury levels in fish had increased approximately a hundredfold from 1935 to 1970. Methyl mercury makes up 95 to 100 percent of the mercury found in fish [52].

Many U.S. industries have discharged inorganic mercury into the water; the danger in this practice was overlooked even after the Minamata disaster because organic methyl mercury had caused the poisonings in Japan. It has now been known for a number of years, however, that inorganic mercury can be transformed in nature into methyl mercury which can then be accumulated by fish and passed on to man in this highly toxic form. Mercury can be methylated by the microorganisms (the bacteria and molds) found in lake or river sediments. This reaction can be carried out by any microorganism that can synthesize vitamin B_{12}, that has a large supply of organic substances as a source of food, and that, of course, has a supply of inorganic mercury. Methyl mercury is not bound by the organic material in sediment, but is easily released into the water where it is soluble and available for consumption [53].

Thus, while many industrial firms have halted or reduced mercury discharges, the great accumulation of inorganic mercury in the sediments of the aquatic environment provides a continuing source of methyl mercury and insures continuing contamination of the food chain.

It has also been suggested that bacteria living on the gills of fish can convert inorganic mercury into organic mercury. In addition, experiments with birds have shown that when they are fed nonmethylated mercury compounds

"Minamata disease" in the people of Minamata Bay in Japan proved to be mercury poisoning from eating fish caught from waters contaminated by a plastics plant which used a mercury compound in its processing. Fifty industrial plants in Japan had been designated as in need of supervision "against possible contamination of river or sea water with methyl mercury," according to the *Japan Times*, December 19, 1968. The Water Quality Council, an advisory council to Wataro Kanno, Director General of the Economic Planning Agency, selected the plants and specifically recommended that "the methyl mercury concentration in river or sea water into which plant wastes are discharged should not exceed 0.01 ppm.". . . The move is aimed at preventing a recurrence of Minamata disease outbreaks which occurred in Kumamoto and Niigata prefectures."

they may produce eggs which contain methyl mercury. It is not yet known if this type of conversion takes place in the human intestine [54].

The classic description of the importance of water to the evolution and maintenance of life is that of Lawrence J. Henderson in *The Fitness of the Environment*, first published in 1913. "If doubts remain," said Henderson, "let a search be made for any other substance which, however slightly, can claim to rival water as the milieu of simple organisms, as the milieu interieur of all liv-

Japan was formerly one of the largest users of mercury fungicides, but alerted by the Minamata episode, the Japanese have almost completely abandoned the use of mercury compounds in rice growing.

ing things, or in any other of the countless physiological functions which it performs either automatically or as a result of adaptation" [55].

Yet water's life-giving and life-protecting qualities can sometimes be overcome, and its capacities to dissolve and transport substances can prove harmful as well as helpful to humans and other forms of life. Human and industrial wastes, as well as the naturally-occurring substances which contribute to life, have entered the hydrological cycle in tremendous quantities. They may be locally concentrated or transported worldwide, and in either case may become part of the food chains that have their base in aquatic organisms; they may be hidden from view and collected in the vast reservoir of ground water or in the bottom sediments of lakes and oceans, storing up trouble for the future; or they may change some quality of the water itself and thus disrupt ecological balance.

[1] Kenneth S. Davis and John Arthur Day, *Water: The Mirror of Science,* (Garden City, New York: Doubleday, Anchor Books, 1961), pp. 133–138.

[2] William C. Ackermann, E. A. Colman, and Harold O. Ogrosky, "Where We Get Our Water: From Ocean to Sky to Lane to Ocean," in *Water, The Yearbook of Agriculture* (Washington, D.C.: U.S. Department of Agriculture, 1955), p. 41.

[3] Virginia Brodine, "The Earth-Atmosphere System," Chapter 2 in *Air Pollution,* (New York: Harcourt Brace Jovanovich, 1973), pp. 30–53.

[4] Justin Frost, "Earth, Air, Water," *Environment,* July–August 1969, Vol. 11, No. 6, 14–33. Except as otherwise noted, this article is the source for all the information which follows on the distribution of DDT.

[5] George M. Woodwell, Paul P. Craig, and Horton A. Johnson, "DDT in the Biosphere: Where Does It Go?," *Science,* December 10, 1971, Vol. 74, 1101–107.

[6] Woodwell, "DDT in the Biosphere," p. 1105 (see [5]).

[7] Bernado Houssay, "Water Metabolism," Chapter 44 of *Human Physiology,* Second Edition (New York: McGraw-Hill, 1955), pp. 449–55.

[8] Lawrence J. Henderson, *The Fitness of the Environment* (New York: Macmillan, 1913). Reissued with an introduction by George Wald (Boston: Beacon Press, 1958).

[9] Davis and Day, *Water,* pp. 34–36 (see [1]).

[10] Sterling B. Hendricks, "Necessary, Convenient, Commonplace," in *Water, The Yearbook of Agriculture,* pp. 11–12 (see [2]).

[11] Davis and Day, *Water,* adapted from Figure 3, p. 28 (see [1]).

[12] Ibid., p. 29.

[13] Ibid., adapted from Figure 3, p. 28.

[14] George W. Brown and James T. Krygier, "Effects of Clear-Cutting on Stream Temperature," *Water Resources Research,* August 1970, Vol. 6, No. 4, 1133–39.

[15] John Cairns, "We're in Hot Water," *Environment,* October 1968, Vol. 10, No. 8, 187.

[16] "Industrial Waste Guide on Thermal Pollution," U.S. Department of the Interior, Federal Water Pollution Control Administration, September 1968, adapted from Table 4, p. 13.

[17] "Thermal Pollution of Cayuga Lake by a Proposed Power Plant," Citizens Committee to Save Cayuga Lake, Ithaca, New York, May 27, 1968.

[18] *Water Quality Criteria,* report of the National Technical Advisory Committee to the Secretary of the Interior, Federal Water Pollution Control Administration, Washington, D.C., April 1, 1968, pp. 42–43.

[19] John R. Clark, "Thermal Pollution and Aquatic Life," *Scientific American*, March 1969, Vol. 220, No. 3, 20–21.

[20] F. F. Fish, "The Return of Blueback Salmon to the Columbia River," *Scientific Monthly*, April 1948, Vol. 66, pp. 283–92.

[21] Richard L. Major and James L. Mighell, "Influence of Rocky Reach Dam and the Temperature of the Okanogan River on the Upstream Migration of Sockeye Salmon," *U.S. Fish and Wildlife Service Fishery Bulletin*, Vol. 66, No. 1, 131–47.

[22] "Thermal Fish Kill," *Sports Fishing Institute Bulletin*, March 1972, No. 232, 7.

[23] Clark, "Thermal Pollution and Aquatic Life," p. 22 (see [19]).

[24] Daniel Merriman, "The Calefaction of a River," *Scientific American*, March 1970, Vol. 222, No. 3, 42–52.

[25] *Water*, p. 33 (see [1]).

[26] G. E. Hutchinson, "Inorganic Ions of Rain Lakes and Rivers," Chapter 8 in *A Treatise in Limnology*, Volume I (New York: John Wiley and Sons, 1967).

[27] *Water Quality Criteria*, pp. 40 and 68 (see [18]).

[28] David Leighton, I. Nusbaum, and Stewart Mulford, "Effects of Waste Discharge from Point Loma Saline Water Conversion Plant on Intertidal Marine Life," *Journal of the Water Pollution Control Federation*, July 1967, Vol. 39, No. 7, 1190–202.

[29] *Environmental Impact of Highway Deicing*, Environmental Protection Agency Water Quality Research, Edison Water Quality Laboratory, Edison, New Jersey, June 1971, pp. 51–52.

[30] Ibid., p. 59.

[31] Robert C. Bubeck, William H. Diment, Bruce L. Deck, Alton L. Baldwin, and Stewart D. Lipton, "Runoff of Deicing Salt: Effect on Irondequoit Bay, Rochester, New York," *Science*, June 11, 1971, Vol. 172, 1128–32.

[32] U.S. Department of the Interior Geological Survey, *Primer on Ground Water* (Washington, D.C.: U.S. Government Printing Office, 1963), p. 22.

[33] "Injection Wells Pose a Potential Threat," *Environmental Science and Technology*, February 1972, Vol. 6, No. 2, 120–22.

[34] David M. Evans and Albert Bradford, "Under the Rug," *Environment*, October 1969, Vol. 11, No. 8, 2–31.

[35] "Groundwater Pollution and Conservation," *Environmental Science and Technology*, March 1972, Vol. 6, No. 3, 213–15.

[36] "Urban Soil Erosion and Sediment Control," U.S. Department of the Interior, Federal Water Pollution Control Administration, May 1970, pp. G2–G3.

[37] Kenneth M. Machenthun and William M. Ingram, *Biological Associated Problems in Freshwater Environments: Their Identification, Investigation and Control,* U.S. Department of the Interior, Federal Water Pollution Control Administration, pp. 19–20.

[38] M. M. Ellis, "Erosion Silt as a Factor in Aquatic Environments," *Ecology,* January 1936, Vol. 17, No. 1, 39.

[39] Machenthun, *Biological Associated Problems in Freshwater Environments,* p. 24 (see [37]).

[40] "Phosphates in Detergents and the Eutrophication of America's Waters," 23rd Report by the Committee on Government Operations, April 14, 1970.

[41] Gene E. Likens, Herbert F. Bormann, and Noye M. Johnson, "Acid Rain," *Environment,* March 1972, Vol. 14, No. 2, 33–40.

[42] *Acid Mine Drainage in Appalachia,* a report by the Appalachian Regional Commission, Washington, D.C. 20235, 1969.

[43] C. C. Collier, R. J. Pickering, and J. J. Musser (eds.), *Influences of Strip Mining on the Hydrologic Environment of Parts of Beaver Creek Basin, Kentucky, 1955–66,* Geological Survey Professional Paper 427–C (Washington, D.C.: U.S. Government Printing Office, 1970).

[44] J. J. Musser and R. J. Pickering, "Geochemistry of Water," in *Influences of Strip Mining,* pp. C14–C46 (see [43]).

[45] J. P. Henley, "Stream Bottom Fauna," in *Influences of Strip Mining,* pp. C46–C49 (see [43]).

[46] J. R. Sheridan, "Fish Population," in *Influences of Strip Mining,* pp. C50–C53 (see [43]).

[47] R. H. Weaver and H. D. Nash, "Microbiology of Streams," in *Influences of Strip Mining,* pp. C53–C57 (see [43]).

[48] R. Lloyd, "The Toxicity of Zinc Sulphate to Rainbow Trout," *Annals of Applied Biology,* Vol. 48, 84–94. Cited in J. F. Skidmore, "Toxicity of Zinc Compounds to Aquatic Animals with Special Reference to Fish," *The Quarterly Review of Biology,* September 1964, Vol. 39, No. 3, 232.

[49] Peter Doudoroff, "Some Recent Developments in the Study of Toxic Industrial Wastes," Proceedings of the 4th Pacific Northwest Industrial Waste Conference, State College of Washington, Pullman, Washington, 1952, pp. 21–25. Cited in Peter Doudoroff and Max Katz, "Industrial Wastes: Critical Review of Literature on the Toxicity of Industrial Wastes and Their Components to Fish. II. The Metals, as Salts," *Journal of the Water Pollution Control Federation,* July 1953, Vol. 25, No. 7, 811.

[50] Neville Grant, "Mercury in Man," *Environment,* May 1971, Vol. 13, No. 4, 9 and 11.

[51] Göran Löfroth with Margaret E. Duffy, "Birds Give Warning," *Environment,* May 1969, Vol. 11, No. 4, 12.

[52] Grant, "*Mercury in Man,*" p. 9 (see [50]).

[53] John M. Wood, "A Progress Report on Mercury," *Environment,* January–February 1972, Vol. 14, No. 1, 33–39.

[54] Grant, "*Mercury in Man,*" p. 9 (see [50]).

[55] Henderson, *The Fitness of the Environment,* pp. 131–32 (see [8]).

Ocean Trial

The major element in the worldwide hydrological cycle is the ocean. In addition to the contaminants dumped directly into ocean waters, dangerous pollutants enter the sea in huge amounts from both air and freshwater pollution. Efforts to control the many sources of marine pollution are fragmentary and are complicated by the lingering attitude that the deep ocean is a no man's land—an inviting open dump that can be closed only by putting international cooperation to the test.

Ocean pollution had attracted little public attention until the curtain rose on a particularly dramatic episode on March 18, 1967. At about 9:00 that morning, the *Torrey Canyon,* then one of the ten largest ships in the world, was cutting through the water at 16 knots off Lands End, England. Between the supertanker and its intended port of Milford Haven lay an infamous chain-of-rocks reef that had claimed many ships. The oncoming tanker, low in the water with a full load of crude oil from Kuwait on the Persian Gulf, was spotted by lookouts on the Seven Stones lightship who determined that it was on a collision course with the reef. The crew of the lightship frantically tried to warn the *Torrey Canyon* with flares, emergency flags, and rockets. The sea was calm and sunlit, and the watch on the tanker had ample time to recognize the danger and change course. The lightship signals went unheeded, however, and, despite good visibility, good weather, and adequate warning, the *Torrey Canyon* ground onto a rock pinnacle with an impact that tore a 500-foot gash through six of its 18 oil storage tanks. The wreck focused world attention on modern marine pollution problems [1].

Giant Tankers—Giant Problems

The loss of the tanker was attributed to negligence on the part of its captain in later maritime hearings [2], but the environmental problems caused by the 95,000 tons of crude oil released from its hulk are part of a worldwide pattern that cannot be understood on the basis of human inattention in one isolated case. The reasons for the rising tide of pollution at sea and in fresh water are more closely related to the underlying trends and assumptions that motivate our current approaches to the transport of material on water, the manufacture of goods, and the disposal of wastes. There were two major incentives for the development of the modern supertanker. One was that the larger the tanker—up to a point—the lower the construction costs per ton of tanker carrying capacity and the lower the operating cost. The other stimulus was that these ships could bypass the Suez Canal, risky in politically troubled times, and still show considerable profit because of the large volume of oil they carried [3].

The commitment to construct giant tankers was originally made primarily on economic grounds, with no serious assessment of the possible environmental hazards. Now, engineering problems and the risk of mammoth oil spills have boosted annual insurance premiums to as much as $1 million for each ship, or more than four times the rate originally anticipated by shippers [4]. Despite this sharp reduction in their profitability, some 200 of the big ships have been ordered during recent years. The vessels are technology's answer

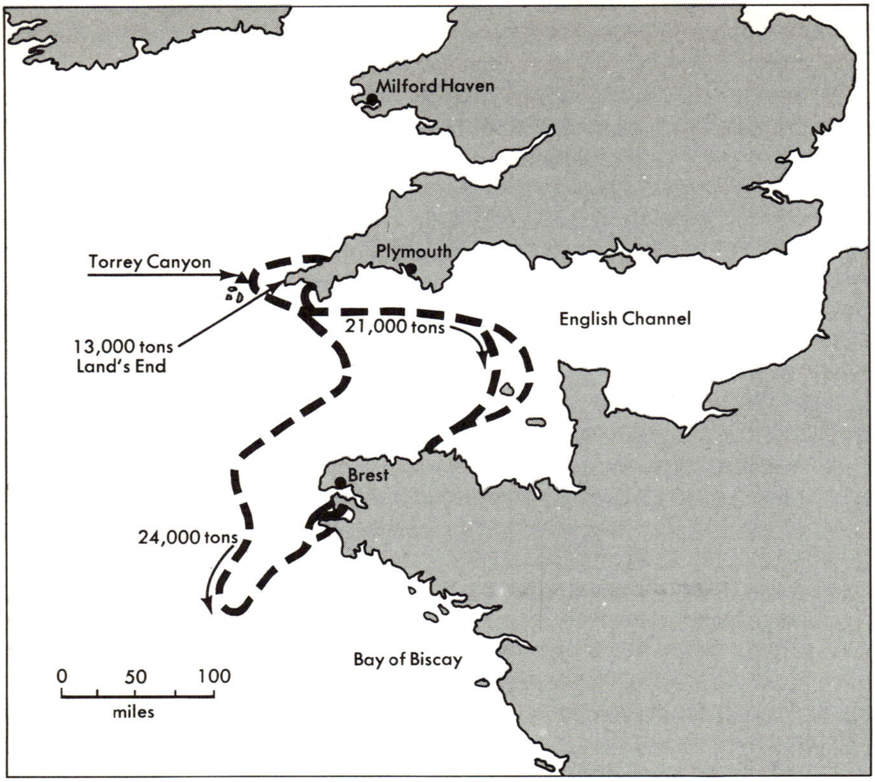

The path of the oil spill from the tanker *Torrey Canyon* which grounded on the Seven Stones reef, 15 miles west of Land's End, England, in 1967.

to the need to move ever larger volumes of crude oil to meet the increased worldwide demand for petroleum products. As a result of this demand, some 1180 million metric tons of crude oil were carried by tanker in 1969. In 1980, more than twice that amount—a total of 2700 million metric tons—is expected to be transported by ship [5].

Although the environmental hazards associated with large tankers were dramatized for shippers after-the-fact by economic constraints such as expensive antipollution insurance, the hazards were predictable before recent accidents. In the first place, the size of the largest ships ordered made them too unwieldy to comply with key provisions of the international rules designed to prevent accidents at sea. New tankers of 400,000 tons deadweight—a measure preferred by oilmen as an indication of cargo capacity—will require a distance of five miles to stop when traveling at full speed. Conversely, they will not maneuver readily at slow speeds. As a result, the ships will not be able to meet international stopping or maneuvering requirements in low visibility or in narrow channels [6]. Smaller supertankers of 200,000–330,000 tons deadweight—the size most common now—also suffer from limited maneuverability

Surrounded by oil from the *Torrey Canyon,* soldiers and firemen try to prevent a slick from moving into Porthleven Harbor, Cornwall, England.

at low speed, although to a lesser degree. This drawback is particularly ominous since at least 80 percent of tanker accidents occur in areas of restricted navigation, such as harbor entrances [7].

At the same time, the trends in tanker losses since World War II indicate increased risks associated with heavier traffic—a situation that cannot be expected to improve with the introduction of even larger, less maneuverable vessels. The number of tankers at sea rose from 2500 in 1955 to 3600 in 1966; by 1983, the number will be an estimated 4300 [8]. In addition, the overall world fleet of tankers, bulk carriers, and freighters is expected to grow from a total of 17,369 in 1966 to 21,468 in 1983. Heavier traffic in already congested shipping lanes has contributed to a rising loss in tanker tonnage at sea. The loss totaled about 200,000 tons in 1948, rose to about 430,000 in 1960, and then increased to more than 550,000 in 1963 [9]. Supertankers have been involved in several incidents over the past few years, the most spectacular of which was the sinking of the 207,000-ton *Marpessa* off West Africa in December 1969. The *Marpessa* was the largest ship ever to sink, so huge that it was equivalent in tons to all of the tankers lost at sea in 1948. Fortunately, the

In recent years, nearly 200 new oil supertankers have been ordered or are under construction. By reducing operating costs and vastly increasing cargo capacity, shipowners foresaw great savings in these new ships. With the rise in conventional tanker accidents and costly oil pollution incidents, however, insurance rates for giant tankers have risen sharply, reducing the expected margin of profit. The new ships generally are built with improved oil-pollution control equipment, but their size increases the possibility of unprecedented pollution disasters in accidents at sea.

vessel was traveling between oil runs and had sea water ballast rather than oil in its cargo tanks. Had the ship been carrying oil, it could have released twice as much as gushed from the *Torrey Canyon* [10].

Less Drama, More Pollution

Although tanker accidents have produced the most vivid oil pollution incidents, other human activities contribute a much greater volume of petroleum hydrocarbons to the ocean. About 90 percent of the total amount of hydrocarbons put into the ocean by man each year comes from normal tanker operations, other ships, refineries, petrochemical plants, and offshore oil wells; from the disposal of spent lubricants and other industrial and automotive oils; and from the fallout or washout of airborne hydrocarbons emitted by industries and motor vehicles.

Human and natural inputs to the ocean of petroleum hydrocarbons from land, sea, and atmosphere in metric tons during 1969 and estimates for 1975 and 1980 are tabulated below [11]. The data disclose that large amounts of hydrocarbons result from routine ship operations and from industrial and automotive wastes. But all direct discharges are overshadowed by input from evaporated or incompletely burned airborne hydrocarbons from chemical factories and from road vehicles. About 90 million metric tons of petroleum hydrocarbons were discharged into the atmosphere in this way in 1969; perhaps ten million metric tons subsequently entered the ocean in rainfall or on dust particles. By 1980, these amounts may double, since petroleum production will more than double by that year. In 1980, then, some 4.7 million metric tons of hydrocarbons may enter the ocean directly from man's ships, industrial plants, vehicles, and ocean oil wells. Up to 20 million metric tons of hydrocarbons originally released from factory and vehicular exhaust will come from the atmosphere.

	1969	1975	1980	Annual Average
Routine Tanker Operations	530,000	805,000	1,062,000	
Other Routine Shipping Operations	500,000	705,000	940,000	
Offshore Wells	100,000	320,000	460,000	
Refinery Operations	300,000	450,000	650,000	
Liquid Oil Wastes (from industry and automobiles)	550,000	825,000	1,200,000	
Accidental Spills (from all sources)	200,000	300,000	440,000	
SUBTOTAL (DIRECT INPUT)	*2,180,000*	*3,405,000*	*4,752,000*	
Airborne Hydrocarbons (from factories and vehicles)	10,000,000	—	20,000,000	
TOTAL (ALL HUMAN SOURCES)	*12,180,000*	—	*24,752,000*	
Natural Underwater Seeps				less than 100,000
Natural Organic Production				3,000,000
TOTAL (NATURAL SOURCES)				*3,100,000*

In comparison with input from human activities, less than 100,000 metric tons of crude oil are estimated to be released annually in the oceans by natural leaks from offshore oil deposits such as those near Coal Oil Point, California [12]. Hydrocarbons produced naturally by marine plants total only three million metric tons each year [13]. Furthermore, crude oil and oil products released by man are persistent, toxic chemicals, whereas natural hydrocarbons released by organic activity are less complex and, with a few exceptions, nontoxic [14]. By 1980, then, the total of 24.7 million tons from the business of man will be eight times the entire natural hydrocarbon output from marine biological activity and from natural underwater seeps—a situation with serious implications for marine life and even for human health.

Despite the predominance of airborne hydrocarbon pollutants, oil pollution control strategy at sea is aimed primarily at reduction of discharges during transport or at cleanup of discharges once they occur. These are necessary measures, but a comprehensive strategy requires identification and control of all sources of pollution throughout the hydrological cycle.

The large input of hydrocarbons from human activities also demonstrates the flaw in the popular retort to environmentalists that nature, not man, is the dominant polluter in the world due to natural processes such as erosion and volcanic eruptions. This sidesteps the problem of toxic pollution, which is caused not by large-scale geological forces but by the discharge of pollutants such as chemicals and metallic compounds in amounts that are large relative to the amounts found naturally in the environment.

As to the metals, the amounts of at least one—lead—appear to be fairly evenly divided between man-made and natural sources, with man-made sources possibly having the edge. According to Dr. Edward D. Goldberg of Scripps Institution of Oceanography:

> In the Northern Hemisphere about 350,000 metric tons of lead, as the anti-knock agent tetraethyllead, is burned in automobile internal combustion engines and subsequently introduced into the atmosphere. . . . About 250,000 metric tons of lead is annually washed out over the oceans and about 100,000 metric tons over the continents in correspondence to their relative areas. Some of the land fallout eventually reaches the oceans as river runoff. This compares with an annual input of lead into the oceans through natural weathering processes of 150,000 metric tons [15].

The concentration of lead, which in excessive amounts can be magnified along the food chain and become hazardous to aquatic life [16], has risen from a prehistoric level of about .01 to .02 ppb to a current level of .07 ppb in surface waters of the oceans in the Northern Hemisphere. The increase is primarily due to the modern combustion of fuels containing lead [17].

The amount of mercury that enters the ocean each year from human activities also may be approaching the amount produced by natural sources, although there is considerable variation among estimates of respective inputs because of the scarcity of investigative data. At least 25,000 metric tons of mercury are washed down in rain into the ocean each year, and the total may be in excess of 100,000 metric tons [18]. The largest amount of mercury in the atmosphere contributing to this washout is of natural origin due to vaporization from the earth's crust. This source may put from 25,000 metric tons of vaporized mercury to four times that amount or more into the atmosphere each year. By comparison, up to 3000 metric tons are discharged annually from chloralkali plant stacks [19], at least 100 tons from cement works, and as much as 20,000 metric tons annually from the burning of fossil fuels [20]. Other amounts enter the ocean as the result of liquid waste discharges to rivers. Because of the wide range in the various estimates that have been made, a precise balance sheet between man-made and natural contributions cannot be made. Clearly, however, the amount attributable to man is considerable, and produces serious environmental and human health threats. Before the modern industrial release of mercury, the contamination of fish by natural methyl mercury was a localized problem. Industrial pollution, however, has vastly magnified this contamination [21]. Other metals with potentially serious implications for marine biology and, in some cases, for human health which now are released by man in quantities rivaling those in nature include copper, nickel, chromium, and cadmium [22].

To Ocean from Air and Surface Waters

The United States alone now releases some 100 million metric tons of particulate materials to the atmosphere each year from the combustion of fuel used for transportation and other needs and from various industrial and agricultural activities. This equals the amount disgorged by volcanoes each year. The amount of particles released to the atmosphere naturally as rock debris and sea salt may total as much as two billion metric tons each year, far more than that now produced by man in the United States [23]. However, the particles released by man, including material such as fluorides, beryllium, lead, and asbestos, can be hazardous to life [24]. Thus man's pollution can surpass nature's in toxicity, if not in quantity. These considerations, plus the need to account for discharges from expanding technologies in other countries of the world, led a panel of the National Academy of Sciences (NAS) to conclude that "a modern industrial society can now compete with natural processes as a source of materials in the environment. Although the available data are poor in quality, they suggest that burning of fossil fuels will have a significant impact on the marine environment." [25]

The impact of man-made versus natural pollution at sea can be seen most sharply in the accumulation of synthetic organic chemicals in the environment that can be detrimental to life. The list of such chemicals is long, but in a 1971 NAS report prepared under the auspices of the Ocean Science Committee, three groups were selected as representative: chlorinated hydrocarbon insecticides used for agricultural and public health, chlorinated hydrocarbons for industrial use, and volatile organic liquids and gases for various applications. The table on page 77 presents representative amounts of chlorinated hydrocarbon insecticides produced and released to the ocean [26]:

Chemical	Annual Global Production (in metric tons)	Annual Amount Reaching the Sea (in metric tons)
DDT	100,000	25,000
Aldrin-Toxaphene (aldrin is rapidly converted to dieldrin)	100,000	25,000
Benzene Hexachloride (BHC)	100,000	more than 50,000

The main route of transport for these insecticides was the atmosphere [27], where the chemicals were carried along on dust particles or as vapor, and subsequently were discharged into the sea in rain, dry fallout, or possibly were taken up directly by surface films of oil that act as solvents for the substances.

The atmosphere was also the major route for the dispersal of PCBs, selected in the NAS report as representative of the chlorinated hydrocarbons used in industry. PCBs are compounded as insulators, transformer oils, heat exchangers, plasticizers, and a variety of other industrial aids. At the time of the 1971 NAS report, an estimated total of about one million metric tons of PCBs were produced in the world; 25,000 of these metric tons ended up in the ocean. PCBs are toxic, as are chemicals used as industrial PCB substitutes such as diphenyl, triphenyl, diphenyl ether, and alkylnapthalenes, according to the NAS report.

The third group of common synthetics reported as potentially harmful environmental contaminants were represented by 1,2 dichloroethane, a volatile liquid used in the manufacture of vinyl chloride, as a gasoline additive, and as a cleaning solvent; freons, used in the gaseous state as propellants in aerosol cans and as refrigerants; and dry-cleaning solvents such as those from petroleum naphtha. Global production estimates were five million metric tons a year for the first compound, 400,000–600,000 metric tons for freons, and two million tons for dry-cleaning solvents. A large amount of the first liquid would be anticipated in the environment because of its volatile nature, but how large an amount could not be ascertained. It was estimated that much of the annual freon output was released to the atmosphere from the discharge of aerosol cans and then carried great distances by global winds. The total atmospheric burden of freons was estimated as 240,000 metric tons, with the average time in the air about one year before deposit in land or sea. Of two million metric tons of dry-cleaning solvents produced in the world, according to the NAS study, perhaps 50–80 percent evaporated to the atmosphere,

An industrial stack, with its plume of particulates reflected in the water, symbolizes the contribution of air pollution to the later pollution of water. The particles have to go somewhere, and many of them end up in the water.

while lesser amounts were discharged into sewers. Thus, the amounts of these solvents that reached the oceans were considerable. Other synthetic, volatile organic compounds such as vinyl chloride, carbon tetrachloride, aldehydes, ketones, and alcohols also were released into the environment and thus were susceptible to transfer to the ocean by precipitation, fallout, or direct dissolution as gases. Global poroduction of these light synthetic compounds was approximated to be 20–30 million tons annually.

The foregoing estimates of amounts of synthetic chemicals released to the environment were based generally on overall production figures and use patterns—the only practical procedure available, but one which provides merely the roughest approximation. Another phase of human activity—the intentional dumping of large quantities of wastes into the sea—produces an additional threat from chemicals that is not reflected in estimates such as the preceding NAS figures. For example, chemical wastes from pesticide factories in the Mississippi River Valley, principally in the Memphis area, were shipped at an estimated rate of 14,000 tons per month in 1968 to a point 100 miles into the Gulf of Mexico and dumped into the water. That year, an overall total of 330,000 tons of pesticide wastes passed into waters off the U.S. coast, according to a landmark study performed for the Solid Waste Management Office by members of the Dillingham Corporation, La Jolla, California [28]. The pesticides included chlorinated hydrocarbon as well as other types of insecticides, herbicides, and fungicides. These pesticide wastes were part of a total of 4.7 million tons of industrial wastes disposed of off the continental United States in 1968 that included 2.7 million tons of acids, 560,000 tons of refinerv wastes, 140,000 tons of paper mill wastes, and 940,000 tons of other assorted materials including ammonium sulfate and sodium cyanide. The majority of these chemical wastes were shipped in tank barges from which the chemicals were discharged directly into the ocean. Some dangerous chemicals were shipped in various types of containers—a practice that protected the working crews rather than the environment, since the containers were generally either purposely ruptured at the ocean's surface or were ruptured by water pressure after they sunk. In any case, the chemical waste usually came into direct contact with the ocean after a period of time.

Few studies have been made on the environmental effect of this waste disposal practice. In one case, industrial chlorinated hydrocarbons in the Gulf of Mexico were found to kill fish and plankton (minute plant and animal organisms) in the immediate disposal area, with effects on marine life lasting two to four hours. Laboratory tests of the chemicals—which included betachloropropylene, trichloropropane, isopropyl chloride, and allyl chloride—revealed inhibition of photosynthesis in tiny marine plants (phytoplankton) and of respiration in very small surface marine animals (zooplankton). The discharge of the chemicals of one tank affected an immediate area of some seven square miles of ocean surface. Perhaps the most glaring shortcoming in the few tests conducted was that no estimates were made of the long-term, cumulative biological impact of the stable chlorinated hydrocarbon compounds and pesticides. As such compounds are concentrated along the food chain, their effects become increasingly serious. Monitoring of the specific types of chemical wastes shipped to sea by state or federal officials has been rudimentary and, in many cases, nonexistent, primarily because of the lack of clear-cut jurisdictional authority. Official control today is particularly difficult, since most industrial wastes are disposed of at least 15 miles from shore—a distance well beyond the three-mile territorial sea over which the United States claims legal authority.

Information about chemical waste disposal off the shores of other countries is sketchy, but data gathered during the late 1960s indicates that the practice in some has already reached serious proportions. West Germany, for example, was disposing of some 40 tons of chlorinated hydrocarbon materials each month in the Atlantic during this period [29]. Other wastes dumped at sea by that country included 375 tons of sulfuric acid each day, 750 tons of iron sulfate a day, 200,000 tons of gypsum per year, and single large discharges of arsenic residues. Ocean discharges from the North Pier, Hook of Holland, contained 1500 tons per year of nitric acid and fatty acids, and 1700 tons per year of metal-pickling bath wastes containing aluminum, iron, chromic chromium, sulfuric acid, hydrochloric acid, and cyanide. English vessels carried a million tons of power plant ash residue and two million tons of coal mine wastes to dumping grounds off the northeast coast of England. In 1970, a team of Norwegian and Swedish scientists reported that a wide stretch of the Atlantic from Greenland to the Arctic Ocean north of Finland bore traces of polychlorinated alipathic hydrocarbons—unwanted byproducts of plastics production that had been dumped at sea off both Europe and the United States [30]. In areas where the chemicals were concentrated, the sea was red with dead plankton, including small crustaceans—important members of the marine food chain. At some of the test areas, where plankton were not killed outright, the photosynthetic capacity of plants was reduced.

The dumping of waste containers, castoff military hardware, and other junk has interfered with fishing in a number of areas of the world, according to a 1971 report from the United Nations Food and Agriculture Organization:

> Wrecked cars and other junk have hampered fishing—particularly in the North Sea and the Baltic—by mechanical damage to nets and boats, and good fishing areas have been closed because of the danger from dumped military wastes such as explosives, cyanide compounds, biological and chemical warfare agents, and radioactive wastes. In the North Sea and along the Norwegian shelf, dumped industrial waste has recently caused great concern to fishermen. One trawler recovered as many as eight containers in one day, about half of which were leaking out their contents of halogenated hydrocarbons and other toxicants. A similar situation has been reported to exist around the Japanese islands [31].

The relentless and often poorly understood effects of tide and current make the dumping of hazardous wastes risky. During the summer of 1969, several incidents occurred in the southern Baltic in which fishermen were injured by World War II mustard gas released from gas bombs they hauled up

These World War I soldiers were protected from the effects of mustard gas—a gas that was used in the first world war, manufactured but not used in the second world war, and dumped in the Baltic Sea after World War II. Baltic fishermen were unprepared for the gas, however; unprotected, they hauled up mustard gas bombs in their nets in 1969, and several were injured.

in their trawling nets [32]. Two of the men required surgical skin transplantations. The bombs were part of some 20,000 tons of chemical warfare material dumped into the Baltic after the war. The dumping site was originally 20 miles north of the area where the fishermen were hurt, but the action of the sea shifted the gas containers into the shallower waters only a few miles from coastal resort areas.

Still another type of pollutant is released by various applications of nuclear energy. Although the U.S. Atomic Energy Commission has greatly restricted the amounts of radioactive material that can be discharged at sea, a total of 86,758 containers with low-level radioactive wastes were sunk in the ocean off U.S. coasts from 1946 through 1970. At the time of disposal, these wastes contained an estimated 94,673 curies of radioactivity [33]. Low-level radioactive wastes also are discharged regularly at sea from the growing fleet of U.S. nuclear submarines. Another source of undetermined scope is the discharges of nuclear power generating plants [34]. The long-distance transfer of radioactive wastes in the ocean is illustrated by evidence that a southbound current along the western coast of the United States may have transferred zinc 65 from the mouth of the Columbia River to tuna fish ultimately caught off San Diego. The radioactive substance probably came from the Hanford plutonium plant, which discharged radioactive zinc into the Columbia River at that time [35]. Most of the Hanford reactors have now been shut down, but radionuclides continue to accumulate in the ocean due to both civilian and military uses of nuclear power. "Individually, none of these sources adds significant amounts of radioactivity to the ocean. Taken together, however, the increases could be of significant concern," concluded the council on Environmental Quality in its 1970 report to the President on ocean dumping [36].

The most thorough study of the effects of ocean dumping has been conducted in the New York Bight, an area of the continental shelf off New York Bay which contains official disposal grounds for a variety of castoffs. A 1970 report by Dr. M. Grant Gross and associates at the State University of New York, Stony Brook, presented estimates of the total amounts of oxidizable carbon—presumed to place a heavy demand on the dissolved oxygen content of the water—and of potentially hazardous metals in the dredging wastes, sewage sludge, and fly ash dumped in the New York Bight. The data presented below [37] are adapted from a table prepared by Dr. Gross on the basis of amounts of wastes put in the Bight during 1964–1968 and of chemical analyses of metal samples taken during the study.

Type of Waste	Annual Discharge (metric tons)	Annual Content Oxidizable Carbon (metric tons)	Annual Content of Certain Metals (metric tons)				
			Silver	Chromium	Copper	Tin	Lead
Dredging Wastes	3,500,000	60,000	50	1400	700	200	2000
Sewage Sludge	200,000	40,000	2	200	200	60	100
Fly Ash	100,000	3,000	less than .1	20	10	less than 2	30

Gross and his investigators pointed out that although the metals listed were known to be toxic, carcinogenic, or both, the chemical form of each element, which was not determined in the tests, would have to be known before their threats could be assessed. The disposal of large amounts of oxidizable carbon in a relatively small area of the ocean, however, posed a clear threat to the dissolved oxygen content of the water, which generally is shallow in the dumping area except for a deep trough—the Hudson Channel—carved across the shelf floor by the outward flow of the Hudson River. The more than

100,000 metric tons of oxygen-demanding carbon disposed of each year as waste were about 19 times the amount that would be produced by natural biological processes in the dumping area, about 35 square nautical miles in size.

To put the scale of the dumping operation in perspective, the annual total of at least 8.6 million tons of solid material in dredging, construction, sewage sludge, and acid wastes dumped in western Long Island Sound and in the New York Bight during 1964–1968 was the largest single sediment source on the eastern coast of the United States then. Put another way, the entire New York metropolitan region produced about 1200 tons of solid wastes per square kilometer of area each year during that period—an amount that exceeded the sediment yield per unit of drainage area in any major North American river basin [38]. This provides another indication that industrialized societies now rearrange large quantities of materials to a degree once reserved for nature.

Marine studies begun in 1968 by scientists at the Sandy Hook Marine Laboratory in New Jersey in a joint study with other agencies demonstrated the extent to which the various wastes in the New York Bight harmed ocean life. The sewage sludge had spread north from the designated dumping ground until it covered an area of more than 14 square miles. The heavier dredging wastes dispersed over a smaller area—about seven square miles. Dissolved oxygen in near-bottom water over the sewage sludge site was frequently below 2 ppm from late July to mid-September, a level not high enough to sustain marine life in laboratory experiments [39]. In the polluted areas, bottom-dwelling marine organisms longer than one millimeter were completely absent or were found only in greatly reduced numbers in comparison with numbers in nearby uncontaminated regions. In the sewage sludge area, not even usually pollution-tolerant marine worms could survive. A study of minute bottom organisms less than one millimeter in length disclosed the same trend: One sample from the center of the dredging spoils site had only nine species; another, only two—a decrease in species diversity from that found elsewhere in the New York Bight [40].

In regard to larger shellfish, the rock crabs found in other areas were seldom discovered in the waste disposal areas. Adult crabs that were in the waste beds frequently were diseased or near death, an indication of the hazard to crabs and lobsters that crossed the waste sites during their seasonal migrations from inshore to offshore water. In laboratory experiments, lobsters and crabs placed in aquaria containing sewage sludge were afflicted with wounds on their shells and their gills became fouled with granular material.

Flounder and ling (bottom-dwelling fish) caught in the vicinity of the sewage sludge disposal area had been feeding on smaller bottom organisms. Stomach analysis showed that the fish also had taken in hair, bandaids, seeds, and cigarette filters from the sludge. Some flounders had black gills, a condition that was reproduced experimentally in healthy fish kept in aquaria with sludge. Levels of nickel, chromium, and lead in several fish exceeded amounts considered normal for marine animals—possibly an indication that metals from the dumping sites had been concentrated in the fish through the food chain.

Another observation was the spread of toxic or disease-bearing substances across the ocean floor. Heavy metals were carried to the north of the disposal sites and as far as 25 nautical miles south of the sewage sludge station. Water from sediments up to six miles from the center of the sewage sludge dump contained coliform bacteria—indicative of fecal contamination at levels that exceeded those recommended for estuary waters used for shellfishing, an enterprise popular in the New York Bight.

Experiences elsewhere in the world parallel conditions in the New York Bight in certain respects. According to the Sandy Hook Laboratory investigators, other studies have shown apparent accumulation of heavy metals in sludge and sewage dumps in the Thames River Estuary, England; the Firth of Clyde, Scotland; and the coastal waters off California. Studies in a Delaware Bay sludge disposal ground, however, have shown no accumulation of heavy metals—nor, for that matter, of sludge. A number of factors may account for variances in the effects of waste at different dump sites. For example, about 53 percent of the sewage sludge dumped in the New York Bight is untreated and therefore has a higher content of fecal matter, other organic material, and, possibly, toxic metallic compounds than would be found in chemically or biologically treated sludge [41]. Sludge from Philadelphia has received biological treatment (digestion) that renders it less hazardous when dumped at sea [42]. Natural elements that cause variations in waste effects include water depth, bottom topography, currents, winds, and tides. Against these variables, however, must be set the invariable threat posed by toxic materials disposed of in water anywhere. Long-lived toxic substances may be swept out of the immediate disposal area only to cause serious biological problems in other waters or in the marine food chain.

To shift back to the national scene, of the total 62 million tons of wastes dumped off the continental United States in 1968, more than 52 million tons were dredging wastes. This was nearly 15 times the amount dumped in the New York Bight, as we can see in the preceding table. The U.S. Corps of Engineers estimated that about 34 percent of these dredging wastes were polluted to some unspecified degree [43]. Even material with no direct toxic effect, however, could have produced sediment buildup or turbidity that interfered with marine life. To round out the picture, in addition to the 52 million tons of dredging wastes and 4.7 million tons of industrial wastes categorized earlier, 4.5 million tons of sewage sludge, 26,000 tons of refuse and garbage, 574,000 tons of construction and demolition debris, 15,200 tons of outdated military explosives and chemicals and small amounts of miscellaneous waste were dumped into U.S. coastal waters in 1968.

The trend in ocean disposal is ever upward. Available data on disposal of wastes exclusive of explosives, radioactive wastes, and dredging materials (which account for the largest amount of ocean wastes) show a fourfold increase in tonnage from 1949 to 1968—up from 1.7 million tons in 1949 to 7.4

million tons in 1968, as averaged over certain periods during that time. In the years 1950–1968 (approximately the same period), world population increased by less than one-and-a-half-fold (from 2.4 billion to 3.483 billion) and world energy consumption—a rough index of industrialization—increased by less than two-and-a-half-fold (from 2610.9 million metric ton coal equivalents to 6306). It appears that shipping wastes to the sea as a method of waste disposal is growing at a more rapid rate than could be accounted for simply by population or industrial growth [44].

Finally, these studies do not provide a picture of the amount of industrial and domestic wastes piped directly into the ocean—a practice which also is increasing in industrialized countries. In addition to the discharge of toxic materials from factories, ever larger volumes of heated cooling water with a potential for degrading marine life are being discharged from electrical power generating plants built on estuaries and coastlines to take advantage of the vast source of cooling water at the ocean's edge [45].

Then these are the sewage pipes. In 1970, some 130 sewage waste pipes from the New York Metropolitan region emptied into the New York Bight, producing a total flow of 1843 million gallons per day—16 percent of which received no treatment at all. In addition, some 23 communities along the New Jersey shore pumped untreated sewage sludge directly into the Bight through pipes extending as far as 1000 feet from shore. On a worldwide basis, several million tons of solid domestic wastes are piped through sewers to the oceans each year [46], an amount that can be expected to increase with the world's population and, particularly, with the continued growth of the world's coastal cities.

Toward More Tranquil Seas . . .

To solve the problem of ocean pollution requires, for one thing, a careful reassessment of human activities in light of the relative influx of waste from various sources. For example, to be effective, commendable programs to reduce oil pollution from tankers by retaining oil wastes on board instead of pumping them into the sea during routine operations and to create international agreements that control oil pollution at sea [47] must be complemented by controls on hydrocarbon wastes put to sea from both airborne and riverborne sources. Considerations for potential contamination from large new chemical tankers must be extended to serious daily contamination from the intentional dumping of large volumes of chemical wastes in the ocean. Expenditures to treat sewage sludge routinely discharged from barges at sea must be applied as well to the treatment or elimination of equally toxic materials piped directly into the ocean from shoreside cities. By the same token, efforts to eliminate the direct discharge of mercury, other heavy metals, and many chemicals into the ocean will be incomplete without control of the airborne contamination from industrial and domestic processes that discharge similar materials. The degree to which these controls should be applied calls for a careful weighing of the benefits obtained from the activities involved against the environmental and health costs those activities create.

Although some general pollution control measures and alternatives will be considered later, the technological and legal details of water pollution control must here give way to the more fundamental consideration of the relationship of human activities to water use—a relationship that must be clearly understood if scientists, technicians, lawyers, politicians, and the public in general are to agree on the weights to be given to the entries on the cost-benefit balance sheet.

[1] John M. Marshall, "The Black Wake of *Torrey Canyon*," *U.S. Naval Institute Proceedings*, December 1967, 7.

[2] Thomas A. Clingan, Jr., "Oil Pollution, No Solution?" *U.S. Naval Institute Proceedings*, May 1969, 63–75.

[3] Julian McCaull, "The Black Tide," *Environment*, November 1969, Vol. 11, No. 9, 2–16.

[4] "Pop Goes the Profit," *The Economist*, February 28, 1970, 78–79.

[5] *Man's Impact on the Global Environment*, Report of the Study of Critical Environmental Problems (SCEP), sponsored by the Massachusetts Institute of Technology (Cambridge: The MIT Press, 1970) p. 266.

[6] Capt. Edward F. Oliver, "Gargantuan Tankers: Privileged or Burdened?" *U.S. Naval Institute Proceedings*, September 1970, 39–45.

[7] McCaull, "The Black Tide" (see [3]).

[8] Litton Systems, Inc., *Oceanborne Shipping: Demand and Technology*, report commissioned by the U.S. Department of Transportation, Part 1, June 1968, pp. 1-1–3-162.

[9] Secretary of the Interior and Secretary of Transportation, "Oil Pollution—A Report to the President" (Washington, D.C.: U.S. Government Printing Office, 1968), pp. 10–11.

[10] "Horizon to Horizon" (staff report), *Environment*, March 1971, Vol. 13, No. 2, 13–21.

[11] Table adapted from data in *Marine Environmental Quality*, report of the Ocean Science Committee, National Academy of Sciences, Washington, D.C., 1971, pp. 6–8; William H. Matthews (ed.), *Man's Impact on Terrestrial and Oceanic Ecosystems* (Cambridge: The MIT Press, 1972), p. 302; and M. Blumer, H. L. Sanders, J. F. Grassle, and G. R. Hampson, "A Small Oil Spill," *Environment*, March 1971, Vol. 13, No. 2, 2–12. (For 1975 and 1980, items 1–6 are maximum estimated amounts, as explained in *Marine Environmental Quality*.)

[12] *Marine Environmental Quality*, pp. 1–107 (see [11]).

[13] *Man's Impact on Terrestrial and Oceanic Ecosystems,* pp. 1–540 (see [11]).

[14] Blumer et al., "A Small Oil Spill" (see [11]).

[15] Edward E. Goldberg, "Chemical Invasion of Ocean by Man," in *Yearbook of Science and Technology,* (New York: McGraw-Hill Book Company, 1970), p. 67.

[16] B. W. Halstead, "Toxicity of Marine Organisms Caused by Pollutants," paper presented at the FAO [Food and Agriculture Organization of the United Nations] Technical Conference on Marine Pollution and its Effects on Living Resources and Fishing, Rome, December 9–18, 1970. "Lead in the Sea," *Marine Pollution Bulletin,* January 1971, Vol. 2, No. 1, 8.

[17] *Marine Environmental Quality,* p. 15 (see [11]).

[18] Herbert V. Weiss and Minoru Koide, "Mercury in a Greenland Ice Sheet: Evidence of Recent Input by Man," *Science,* November 12, 1971, Vol. 174, 692–94.

[19] Ibid., p. 694.

[20] Paul R. Ehrlich and John P. Holdren, "Review: The Closing Circle," *Environment,* April 1972, Vol. 14, No. 3, 33.

[21] John M. Wood, "A Progress Report on Mercury," *Environment,* January-February 1972, Vol. 14, No. 1, 33–39.

[22] *Marine Environmental Quality,* p. 12 (see [11]).

[23] *Chlorinated Hydrocarbons in the Marine Environment,* report of the Panel on Monitoring Persistent Pesticides in the Marine Environment, Committee on Oceanography, National Academy of Sciences, Washington, D.C., 1971, p. 8.

[24] *Air Quality Criteria for Particulate Matter,* U.S. Department of Health, Education and Welfare, NAPCA Pub. AP-49, January 1969.

[25] *Marine Environmental Quality,* p. 38 (see [11]).

[26] *Marine Environmental Quality,* p. 12 (see [11]).

[27] Justin Frost, "Earth, Air, Water," *Environment,* July-August 1969, Vol. 11, No. 6, 14.

[28] David D. Smith and Robert P. Brown, *Ocean Disposal of Barge-Delivered Liquid and Solid Wastes from U.S. Coastal Cities,* written for Solid Waste Management Office, Environmental Protection Agency (Washington, D.C.: U.S. Government Printing Office, 1971).

[29] *Report of the ICES Working Group on Pollution of the North Sea,* International Council for the Exploration of the Sea, Charlottenlund Slot, DK-2920, Charlottenlund, Denmark, July 1969, pp. 11–17.

[30] Tony Loftas, "The New Marine Poison," *New Scientist and Science Journal,* July 29, 1971, 266–67.

[31] *Pollution: An International Problem for Fisheries,* Food and Agriculture Organization of the United Nations, Rome, 1971, p. 40.

[32] "Recent Pollution Incidents, Mustard Gas in the Baltic Sea, July-August 1969," *Marine Pollution Bulletin,* September 1969, No. 15, 17–18.

[33] *Ocean Dumping, A National Policy,* a report to the President prepared by the Council on Environmental Quality (Washington, D.C.: U.S. Government Printing Office, October 1970), p. 7.

[34] Smith and Brown, *Ocean Disposal of Barge-Delivered Liquid and Solid Wastes* pp. 4, 15–20, 34–37, (see [28]).

[35] American Chemical Society, press release reporting on the work of Theodore R. Folsom, Scripps Institution of Oceanography, La Jolla, California.

[36] *Ocean Dumping,* p. 11 (see [33]).

[37] M. Grant Gross, *Analyses of Dredged Wastes, Fly Ash, and Waste Chemicals—New York Metropolitan Region,* Marine Sciences Research Center, State University of New York, Stony Brook, October 1970, adapted from Table 8, p. 30.

[38] M. Grant Gross, "New York Metropolitan Region—A Major Sediment Source," *Water Resources Research,* June 1970, Vol. 6, No. 3, 927–31.

[39] *The Effects of Waste Disposal in the New York Bight—Interim Report for January 1, 1970,* The Sandy Hook Marine Laboratory, U.S. Bureau of Sport Fisheries and Wildlife, Highlands, New Jersey, December 3, 1969.

[40] *The Effects of Waste Disposal in the New York Bight—Summary Final Report,* National Marine Fisheries Service, Middle Atalntic Coastal Fisheries Center, Sandy Hook Laboratory, Highlands, New Jersey, April 1972, p. 13.

[41] "Supervisor of the Harbor Permit Program for Waste Disposal in the Atlantic Ocean, Draft Environmental Statement," U.S. Army Engineer District, New York, November 2, 1971, p. 4. (Available as PB 204 264–D, National Technical Information Service, 5285 Port Royal Road, Springfield, Virginia 22151.)

[42] "Defending Deep-Sea Disposal," *Chemical Week,* October 13, 1971, 65–66.

[43] Smith and Brown, *Ocean Disposal of Barge-Delivered Liquid and Solid Wastes,* p. 22 (see [28]).

[44] Michael Corr, personal communication. The world population figures are from the UNESCO Statistical Yearbooks for 1951 and 1970. Raw data on energy from Joel Darmstadter, *Energy in the World Economy, Resources for the Future* (Baltimore: Johns Hopkins Press, 1971), p. 10.

[45] *Terrestrial and Oceanic Ecosystems,* p. 71 (see [11]).

[46] Goldberg, "Chemical Invasion of Ocean by Man," p. 70 (see [15]).

[47] McCaull, "The Black Tide," p. 2–16 (see [3]).

Thirsty Technology

The conflict between our dependence upon the life-supporting qualities of water and the mounting pressure placed on water resources by waste disposal activites is more sharply defined when we turn from ocean to fresh water pollution. This conflict can be seen more clearly if we consider the popular elementary school comparison of the human body to a factory. Both the body and the factory process certain raw materials in order to create usable products—nutrients to be consumed elsewhere in the body or manufactured goods to be consumed elsewhere in the commercial system. As young school children also learn, both the body and the factory excrete unusable portions of their respective raw materials as waste. Unfortunately, usually teachers conveniently end the analogy here, with the wastes left to fend for themselves. A discussion of waste products is often regarded as unimportant, when, in fact, wastes should be at the heart of the lesson. Frequently, wastes from both the factory and the body end up in the water, which is one of the crucial raw materials that is then taken back into both structures, contaminated with impurities that interfere with essential production processes. The difficulty is that whereas pollution controls can be devised relatively quickly to remove the impurities from water for industrial use, the capacity of the human body and of other living things to remove wastes has developed over evolutionary time, and these laboriously contructed bodily defense mechanisms can be overwhelmed by industrial wastes released in quantities or in chemical forms that are new to nature [1]. This point in the analogy illustrates the crucial difference between life and the industrial technology created by that life.

Industrial and Municipal Contamination

In industry, water is a raw material that is used as an ingredient with other raw materials to create a finished product, as a transporting medium, as a cleansing agent, as a coolant, and as a source of steam in heating and power production. In the chemical industry, which creates extremely dangerous water pollutants, water is the raw material most frequently used in the manufacture of 150 important industrial chemicals, and air is the second most frequently used [2]. Various ores and minerals usually thought of as the most important raw materials in chemical production actually fall much farther down the list than water and air in terms of frequency of use. The volumes of water used by industry and agriculture in the United States far surpass the amount of water used for domestic purposes. The overall withdrawal of water in millions of gallons per day for various uses in 1960, compared with projections for 1980 and 2020 based on current estimates of economic and population growth, is tabulated on page 90 [3]. The data show an ever increasing use of water in all categories except agriculture, where the slight decline reflects anticipated improvement in irrigation efficiency.

The dilemma though is that some of the water essential to these activities is discharged laden with pollutants that interfere with other activities. Industries usually are located near rivers that answer specific water quality and water volume needs. The plants utilize a variety of treatment equipment, if necessary, to maintain desired quality. The more serious problem is the inter-

	Agriculture	Mining	Manufac-turing	Steam-Electric	Municipal (domestic)	Total Withdrawals
1960	138,586	2739	22,926	65,988	20,074	250,315
1980	134,883	6982	41,193	179,469	39,394	401,918
2020	131,662	36,546	98,059	528,949	95,289	890,504

ference with biological processes, including those essential to human life, which have limited self-defense mechanisms against dangerous pollution. The scope of that pollution in the United States can be gauged by the results of the first nationwide assessment of river pollution conducted by federal officials in 1970. Pollution was defined as a breach of various state and federal standards and criteria for the physical or chemical condition of fresh water. On the basis of field tests throughout the nation, 32.6 percent—almost one-third—of all U.S. stream miles were characterized as persistently polluted [4]. The primary human activities causing the pollution were ranked in categories according to their relationship to stream miles polluted and degree of pollution.

According to the federal study, the first-place polluters of U.S. waterways were industries. Ranging from chemical to machinery manufactuers, industries that discharged directly into the water produced an estimated 34 percent of the stream polution in the United States. As one measure of the pollution of industry, its organic wastes in 1968 were sufficient, before treatment, to consume 297 billion pounds of oxygen during decompostion in water as measured by the five-day biological oxygen demand (BOD) test—standard assay used as a general, if imperfect indicator of the strength of organic wastes. (The significance of BOD will be discussed in "Overfed.") Expressed in shorthand, these 297 billion pounds of BOD were nearly twice the amount produced by industry just 11 years before in 1975. The 1968 amount was also more than three times the 8.5 billion BOD pounds of domestic wastes produced the same year in municipalities served by sewers. The combined total of 38.2 billion BOD pounds of organic wastes from industrial and domestic sources in 1968 received various degrees of treatment that reduced the amount actually discharged to U.S. waterways to 13,560 billion BOD pounds—a sharp decrease attributable to improvements in waste treatment facilties during the 1960s.

Municipal wastes ranked a close second to industrial contaminants as river polluters in the federal study. One reason was that the category of municipal wastes included all wastes transmitted through community systems of sanitary sewers. Although the municipal treatment plant is popularly assumed to deal mainly with domestic wastes, industries have increasingly turned to these plants for treatment of industrial effluent and, as a result, the respective volumes of industrial and domestic wastes treated were about equal in U.S. municipal waste treatment plants in 1968 [4,5]. The pollution impact of the waste stream from municipal plants, then, was heavily influenced by indus-

Industries are the first-place polluters of U.S. waterways. Ranging from chemical to machinery manufacturers, industries that discharge directly into the water produced 34 percent of the stream pollution in this country in 1968.

trial wastes, although the degree of influence could not be estimated precisely because of a lack of definitive data on amounts of industrial materials piped in municipal sewers. Another reason municipal wastes ranked high as pollutants was that the excessively high bacterial levels associated with human sewage were a frequent cause of pollution and these levels historically have been the major concern of water pollution control officials charged with protecting human health. This concern is important, but has been allowed to obscure the more recent health hazards posed by industrial chemicals. Still another reason for the prominence of municipal domestic wastes in the contamination of waterways was their heavy load of phosphorus compounds which have been linked with eutrophication in many areas. Amounts totaling 467.1 million pounds of phosphates were discharged in 1968, up 248.4 million pounds from 1957 despite the removal of progessively larger quantities during this period by advanced treatment at municipal sewage plants.

Estimated to cause almost 20 percent of all river pollution were agricultural activities that produced siltation, organic and inorganic nutrients from livestock, pesticide and herbicide runoff, and salinity. These conditions were actually the predominant causes of river pollution in the Northern and Southern Plains states west of the Mississippi River. Another six percent of the pollution in U.S. rivers was attributed to other causes, such as sedimentation due to construction and dredging. About five percent of the pollution was caused by mining procedures that dumped silt, acid, and brine into waterways—primarily in the Appalachian coal mining region where mine drainage has polluted some 10,500 miles of streams in eight states [6]. Lesser amounts of pollution were caused by accidental spills, urban wastes such as street litter, and discharges from electrical generating plants that included hot water, radioactive material, and airborne wastes.

Modern Technology and Increased Pollution

In order to relate these data more closely to the nature of the human activity causing the pollution, EPA analysts compared the rate of population growth during 1964–1968 with the increase in annual production (before treatment) of BOD. They estimated that if the rise in BOD levels in domestic sewage had been directly proportionate to population growth, the 1968 figure would have been only 330 million pounds more than in 1964. Instead, there was an increase of 900 million pounds of BOD from 1964–1968. Similarly, expansion of industrial output to accommodate increased population at precisely the level and composition of per capita consumption in 1964 would have added about 990 million pounds to the BOD production figure instead of the much larger increase of 7.2 *billion* pounds that actually occurred in 1968. Furthermore, population growth alone would have accounted for a rise of only 23 million pounds in the annual total of phosphorus compounds in sanitary sewage instead of the 122.3 million pound increase reflected in the 1968 total. Thus, these analyses showed that only slightly more than 16 percent of the total increase of 8110 million pounds in the annual domestic and industrial BOD production from 1964 through 1968 could be attributed directly to population increase and that only about 19 percent of the phosphorus waste increase could be so attributed. Part of the increase was more apparent than real, since these figures did not include waste from towns and rural areas not serviced by sewers and the proportion of the population receiving this service rose between 1964 and 1968. This factor was responsible for a seven percent increase in total BOD and more than a 27 percent increase in phosphorus wastes over what would have taken place had new sewer building merely matched population growth.

Large balances of the increases in both BOD and phosphorus, were still unaccounted for, however. EPA analysts examined the relationship between pollution and the consumption of goods—an aspect of the problem stemming almost entirely from industrial rather than domestic wastes—to determine the sources of the remaining increases:

> Almost 77 percent of incremental BOD production and 53 percent of the increased discharge of phosphorus to sewers can be traced to the amount and composition of rising consumption of goods by Americans. Significantly, much of that production cannot be considered to have improved the real economic well-being of consumers. Twenty-three percent of the total increase in BOD occurred as a result of the growth of pulp and paper output, where more elaborate packaging has provided much of the impetus for growth. Similarly, no less than 55 percent of the larger output of BOD [that is, the total BOD increase] arose from chemicals production, and an indeterminate but large portion of that increase must be ascribed to [the] expanding use of various disposable products. In the same general way, an estimated 42 million pounds of sewered phosphorus can be ascribed to increased utilization of phosphorus in detergent formulations—an increase in unit use that was again reinforced after 1968 with the appearance of phosphorus-rich "enzyme" pre-soaks and detergent compounds. [7]

EPA analysts were of the opinion that "The economy of the United States has been marked not only by a voracious absolute demand for more goods, but by a relative preference for goods whose production involves a substantial wasting of organic materials to water." [8] Other information shows that this demand is reflected in a number of industrial operations ranging from the basic process of steelmaking to advanced techniques for manufacturing complex chemicals. The emphasis on increased production speed and lighter unit weight in the steelmaking industry means that advanced technology now being introduced will actually produce more of certain types of pollution per ingot ton of steel than did the older technology.

Steelmaking. For example, the basic oxygen furnace, which utilizes pressurized oxygen to speed steelmaking, outproduces the older open hearth furnace, but in so doing requires larger volumes of water to clean and cool combustion gases. These wastewaters contain extremely fine suspended particles which are especially difficult to remove. Continuous casting, in which molten steel is converted directly into semifinished shapes rather than into ingot molds, is gaining favor because it eliminates certain interim steps in the conventional steelmaking procedure, thereby increasing production rates. However, continuous casting also requires copious amounts of water to cool molds and castings, producing as a side effect large quantities of contaminated wastewater. In another modern operation, hot-strip mills with reheating furnaces and continuous rolling procedures can turn a thick slab of steel into a thin sheet about six feet wide and a quarter of a mile long in three minutes or less. This high-speed process answers the demand for large quantities of thin-gauge metal, but it also requires large amounts of water to carry away scale removed from the metal surface and to cool the hot steel sheets. Some of the newer hot-strip mills produce up to 100,000 gallons of wastewater per minute containing scale particles too fine to be removed in ordinary gravity-settling

This hot strip mill, designed to roll more steel at greater speed, exemplifies the emphasis on increased production. Advanced technology produces more of certain types of pollution than did the older technology.

Despite improved treatment facilities like those at Bethlehem Corporation's plant in Burns Harbor, Indiana, new technology in steelmaking creates 30 percent more wastewater volume per ingot ton than the old technology.

tanks. The magnitude of this discharge alone intensifies the waste treatment problem, since even a waste stream flowing at 10,000–25,000 gallons-per-minute is very difficult to treat. As a result of these trends, the old technology utilizing conventional open hearth furnaces, ingot molding of steel, and slower steelrolling procedures produces less suspended solids, cyanides, free acids, combined acids, emulsions, and soluble metals per ingot ton. The old technology also creates only 70 percent of the wastewater volume per ingot ton that is produced by the advanced technology. The larger wastewater volume from the new technology makes wastewater treatment considerably more difficult.

Steelmaking plants using the old technology produce slightly more phe-

	Pounds of Waste Produced per Day		Pounds of Waste Produced per Day per Ingot Ton	
	Old Technology*	Advanced Technology**	Old Technology	Advanced Technology
Suspended Solids	198,514	1,513,610	103.0	184.0
Phenols	131	525	0.069	0.064
Cyanides	56	257	0.029	0.031
Fluorides	63	252	0.033	0.031
Ammonia	161	645	0.082	0.078
Lube Oils	5890	19,500	3.08	2.37
Sulfuric Acid	5810	23,240	3.03	2.83
Ferrous Sulfate	21,640	86,450	11.3	10.5
Emulsions	637	9630	0.332	1.17
Chromium	†	518	†	0.063
Zinc	†	21	†	0.0025
Tin	†	132	†	0.016
Ferric Chloride	†	17,290	†	2.10
Hydrochloric Acid	†	4660	†	0.565
Free Acids	††	††	3.03	3.40
Combined Acids	††	††	11.3	12.6
Wastewater Volume	19 million gallons per day	113 million gallons per day	9860 gallons per day per ingot ton	13,750 gallons per day per ingot ton

*Integrated mill producing 1915 ingot tons each day.
**Integrated mill producing 8230 ingot tons each day.
†Considered to be zero.
††Not recorded.

nols, fluorides, ammonia, and lubricating oil wastes than do plants using the new technology, but in terms of the overall effect of total wastes produced per day, considerably more pollution is created by a typical advanced technology mill than by a typical old technology mill. The former produces more than four times the total ingot tons per day and, in doing so, creates vastly more waste, as can be seen in the table (page 95) presenting typical waste loads of old and advanced steel plants [9].

The comparison between the two mills is indirect, since the output of the mill with advanced processes is much greater. When considering the table, also bear in mind that in both cases the totals are for wastes produced *before* various forms of industrial waste treatment such as sedimentation, coagulation, biological treatment, and sludge lagooning. Furthermore, advanced steelmaking technology is anticipated in only 47 percent of all steel plants by 1977, with 24 percent still using the old technology and 29 percent some form of intermediate technology. The point is that intensified demand for steel has prompted the introduction of new steelmaking methods that inherently produce more waste in a number of important categories [10]. To counter this increased pollution potential, wastewater treatment facilities worth $865 million in replacement value were installed in steel plants by 1966. This represented only 1.44 percent of total investment in steel plants and equipment that year, however. The projection of waste pollution loads to be discharged to waterways in the future from steel mills shows that the anticipated improvements in treatment still will not turn back the waste levels—except those of coke plant chemicals and acid wastes—to any great extent. This is indicated in the following comparison of wastes discharged from U.S. steel mills to waterways in 1963 and those expected to be discharged in 1977 (in millions of pounds *after* waste treatment) [11].

	Wastes Discharged from Steel Mills in 1963	Projected Discharge from Steel Mills in 1977
Suspended Solids	957.0	1317.0
Lube Oils	211.9	233.2
Free and Combined Acids	1010.2	560.2
Soluble Metals	5.57	7.42
Emulsions	19.2	36.6
Coke Plant Chemicals	6.68	2.75
Fluorides	3.54	5.80

In interpreting this table, bear in mind that there may be secondary pollution from waste treatment processes such as biological treatment and sedimentation, a problem to be discussed later in this chapter. Also of impor-

Mechanical picking and chemical peeling are two of the technological changes in the food industry that have added to the organic pollution of water. Mass production and new technology produce neat packages of convenient food for the consumer, but at a price in the degradation of water.

tance, present and planned reductions in acid waste discharges include a much greater reliance on deep-well injection of wastes—a disposal method it is anticipated will be used in 15 percent of steel plants by 1977, as opposed to only two percent in 1963. As already discussed in Chapter 1, this technique does not remove wastes from the environment and may eventually produce pollution problems of its own, however.

Canning and Freezing Foods. The canned and frozen fruit and vegetable industry is another area in which modern technology and increased production promote rather than lessen pollution [12]. Mechanical picking results in products that contain more soil and foreign materials than those obtained by hand collection. At the same time, higher specifications for product appearance have prompted increased use of pesticides that contaminate fruit [13], as well as more complete cleaning processes which, in turn, produce greater volumes of waste water. Chemical peeling also creates greater pollution. Formerly, soft fruits and vegetables such as tomatoes were first scaled by steam or hot water and then peeled by hand. In most modern plants, however, tomatoes, peaches, and other fruits usually are peeled by hot-lye immersion or spray. Despite recirculation to conserve the lye, the highly alkaline peeling solution pollutes waterways and carries along with it dissolved organic material from the produce.

Another trend with mixed blessings in this industry has been its increased discharge of wastes to public sewer systems to take advantage of municipal waste treatment facilities. This trend has decreased direct discharges to waterways but increased certain municipal treatment problems. About 50 percent of canning wastes were discharged to sewers in 1950; this will rise to an estimated 69 percent by 1977. Some 20 percent of wastes from the frozen food industry entered the sewers in 1950, compared with a projected estimate of 35 percent in 1977. Wastewater volumes also have increased in both cases, so that the absolute waste quantity has risen as well as the percentage pumped into the sewers. In many areas, the large seasonal influx of wastes from these industries has exceeded the treatment capacity of public waste treatment plants. This has been a particularly serious problem in California where the fruit and vegetable industries are highly concentrated and where summer low-flow rates in waterways coincide with heavy waste discharges from the plants. Furthermore, sudden influxes of large quantities of pollutants from lye-peeling or blanching operations carry enough alkaline wastes to disrupt some biological and chemical processes at the treatment plants.

The big jump in waste loads sent to waterways from the fruit and vegetable industries can be seen in the following comparison between waste discharges from U.S. fruit and vegetable industries to waterways in 1963 and those expected in 1977 (in millions of pounds *after* industrial or municipal treatment) [14].

	Biological Oxygen Demand (BOD)	Suspended Solids	Total Dissolved Solids
Canned Fruits and Vegetables			
1963	370	425	395
1977	490	565	525
Frozen Fruits and Vegetables			
1963	290	325	315
1977	415	470	455

Invisible Pollution

Large-volume waste discharges discolored by iron compounds or even by fruit and vegetable cuttings make for acutely visible pollution in U.S. waterways, but unseen chemical compunds from other human activities are also serious pollutants. A 1970 literature survey by Arthur D. Little, Inc. [15], revealed that a total of 66 organic chemicals had been in fresh water throughout the world. Some 430 other organic chemicals were suspected to be in the water because of their use in industrial, agricultural, or domestic processes likely to produce waste water, the Little study disclosed. Many of these 496 chemicals are known to have immediate or chronic toxic effects, or are capable of inducing cancer in animal or man. The major sources of these substances are industrial enterprises including, in descending order of importance, chemical plants, petroleum plants, plastic manufacturing operations, tar and gas plants, rubber factories, dye and tanning plants, ore processing plants, textile factories, pulp mills, and pharmaceutical plants. Agriculture contributed chemical pesticides, and domestic life added detergents.

The list of 496 organic chemicals found or suspected in waterways is by no means complete. One hindrance to a comprehensive assessment of modern chemcial pollution has been the historical empahsis on conventional tests for BOD levels and concentrations of suspended and dissolved solids which provide a rough measure of water quality but do not identify specific chemicals that may be serious hazards to life even in small quantities. This problem was stressed in a study reported in 1972 by Dr. Lawrence H. Keith of the National Environment Research Center, Athens, Georgia [16]. Dr. Keith and other EPA investigators studied the chemical compostion of samples from waste discharges from seven plants: two petrochemical plants, one synthetic rubber plant, two petroleum refineries, and two chemical companies. The investigators used gas chromatography–mass spectrometry to obtain what amounted to chemical fingerprints indicative of individual chemicals or classes of chemicals. A total of 33 compounds were identified, only seven of which were included among the 496 discussed in the Little study. Although a number of the 33 chemicals had toxic properties, their impact on the environment was uncertain. Nevertheless, EPA investigators concluded that the daily discharge of more than 300 pounds of tetrachloroethane and trichloroethane from one of the chemical companies might have chronic toxic effects on marine life in a nearby bayou. Polycyclic aromatics, such as indenes and naphthalenes from one of the petrochemical companies, might produce taste and odor problems in drinking water. The daily waste output of 600 pounds of alcohols from the other petrochemical plant were sufficient to cause possible oxygen depletion in another small bayou. Total daily discharges from one refinery were 800 pounds of phenols and cresols, 50 pounds of naphthalenes, and more than 700 pounds of aliphatic hydrocarbons. One of the chemical companies discharged a daily total of 335 pounds of two long-lived chlorinated hydrocarbons—1,1,2-trichloroethane and 1,1,2,2-tetrachloroethane—the latter a particularly toxic compound. The primitive state of our insight into the chemical pollution of waterways is underscored by the fact that the compounds identified from one petrochemical plant, one chemical company, and the synthetic rubber company could not have been predicted on the basis of the raw materials and products listed by the manufacturers. Thus, such lists cannot be relied upon exclusively to identify organic pollutants in industrial effluents.

Just as investigative techniques are outdated, waste treatment techniques in the chemical industries are often conventional ones that actually are more appropriate for sewage. As Dr. Keith pointed out, "Although the industrial wastes differ markedly from domestic sewage in compostion, concentration, physical characteristics, and toxicity, the same types of treatment are often used for both. However, industrial wastes frequently contain organic compounds that are resistant to the treatment applied to domestic sewage, and gross pollution measurements are even less adequate with these compounds than they are with sewage." [17]

Chemical Treatment of Organic Wastes

It appears that environmental problems due to organic chemical contamination will increase in the immediate future. One major factor influencing this trend is the versatility of modern industrial chemistry based on petroleum as a raw material. Formerly, the base materials in the production of aliphatic compounds were natural products such as carbohydrates and fatty oils and coal was the main basis of the aromatic compounds. None of these base materials permitted the variety and volume of products that have become possible with the development of the petrochemical industry.

Petroleum. Almost no petroleum was used as chemical base stock in 1930, according to Dr. Neil A. Iliff, member of the British Royal Commission on Environmental Pollution [18]. By 1950, petroleum was the basis of 40 percent, by weight, of all organic chemical production. This figure had risen to 88 percent by 1970 and, by 1985, it is estimated that as much as 98 percent of all organic chemical production will be based on petroleum. The range of petroleum distillates available coupled with skillful synthesizing techniques have spurred an immense increase in the volume and types of organic chemicals produced over the past 20 years. We have already seen that potentially dangerous chemicals escape in waste water discharged during manufacturing operations in chemical plants and refineries. In addition, some of the final chemical products such as solvents and pesticides escape or are released to the environement and, as discussed previously, may ultimately end up in the ocean. On a global basis, each year up to 20 million tons of such chemicals may enter the environment out of approximatley 60 million tons of organic chemicals produced. About half of the 20 million tons released into the environment are solvents such as methanol, ethanol, higher alcohols and their derivatives, glycol ethers, chlorinated solvents, chlorofluorocarbons, and aromatic solvents. Other chemicals include detergents, organic pesticides, gaseous base chemicals such as olefins and acetylene, and many other chemicals not yet identified. If present trends continue, the amount of these discharged chemicals—many of which have biological effects—will rise relentlessly in the coming years as the 1970 world production of 63 million tons of organic chemicals quadruples to a projected 250 million tons in 1985.

Tanning. In certain other major polluting industries in the United States, the picture is less bleak for a variety of different reasons. In some cases, new economic or consumer demands have prompted changes in production techniques that produce less rather than more pollution. In the tanning and finishing of leather, for example, industrial growth has been limited by both foreign competition and plastic and rubber substitutes for leather [19]. A trend toward reduced waste discharges has been attributed, in part, to the increased use of centralized tanneries in large cities rather than those in remote country locations used in the nineteenth century. The city tanneries are nearer the slaughterhouses, their source of hides, but an added advantage is the proximity of municipal waste treatment plants to which organic tannery wastes can be routed. An estimated 80 percent of tannery wastes are now discharged to municipal sewers as compared with 60 percent in 1950. Another reason for this decreased waste load is the trend toward synthetic substitutes for sole leather, which is tanned by older methods that produce large organic waste loads. One other factor that may have an indirect effect on tannery pollution in the years ahead is the price of hair that can be recovered from the hides and sold as a by-product. If the hair is not removed mechanically for this use, it is pulped chemically by a technique that creates wastes with high contents of lime, sodium sulfide, and dissolved proteins.

These economic considerations, then, rather than less polluting technology will determine future waste loads from the leather tanning and finishing processes. According to government investigators: "The important generalizations are that the tanning industy has not made great improvement in reducing pollution produced per unit of hide processed; modern tanning machinery is not designed with wasteload reduction in mind; and no real breakthroughs are in sight." [20]

Allowing for the economic trends in industrial activity, some additional inplant pollution control, and increased use of municipal waste treatment facilities, projected tannery wasteloads discharged to waterways are expected to *decrease* by 1977 in comparison with 1963 base values (in millions of pounds after industrial or municipal treatment) as shown below [21].

Tannery Wastes	BOD	Suspended Solids	Total Dissolved Solids
Discharged in 1963	49	87	403
Projected Discharge in 1977	39	59	384

Waste Controls

Although changing economic conditions exert the dominant, if indirect, influence on the degree of industrial pollution in some industries, the effort to utilize waste controls is an important factor in others. This effort seems most successful in industries which produce biodegradable organic wastes that naturally lend themselves to conventional waste treatment techniques featuring removal of BOD potential, and less successful in industries such as chemical manufacturers which discharge wastes containing durable, potentially harmful, synthetic organic materials. As detailed elsewhere, however, reduction of protein-containing biodegradable organic wastes by conventional biological treatment frequently creates inorganic nutrient supplies that foster another type of environmental problem—eutrophication. This must be kept in mind in assessing the reduction of BOD wastes in industries that have large organic waste loads.

Dairy Industry. One such enterprise is the dairy products industry, in which large, technologically advanced plants have equipment that, in a happy coincidence, improves both business and waste control [22]. Preparation of butter, cheese, milk, and ice cream involves less product loss to the wastewater stream in large, advanced plants than in old ones. This reduction provides both a considerable financial saving and a marked decrease in organic wastes. Stainless steel piping systems, automatic cleaning devices, and other forms of automation in the new plants improve sanitation. This, in turn, increases the shelf life of dairy products and can, if regulated properly, decrease the amount of soap and chemicals used for cleaning. At the same time, nearly all dairy plants located in cities now discharge waste water to municipal sewers. Rural plants increasingly are using waste treatment techniques that include spray irrigation and aerated lagoons in which biodegradation takes place.

Cotton and Wool Industries. In the cotton and wool industries, scouring, cotton fiber desizing, and washing are the major sources of organic pollution, contributing 50 to 75 percent of the total BOD and solids waste load from a given plant [23]. Lesser amounts of BOD but more serious quantities of toxic materials are added by bleaching wastes and by the acids, salts, and dyes used in dyeing and printing. Removal of the BOD potential has been improved in the wool textile weaving and finishing industry by additional biological and chemical treatment of wastes in the plants and by a rise in dischages to municipal sewers, which is expected to reach 75 percent of the total cotton and wool industrial waste load in 1977 as compared with 34 percent in 1950. Emphasis in the cotton textile finishing industry apparently has been on activated sludge biological treatment or the aerated lagoon. Some 20 percent of that industry's wastes were discharged to municipal sewers in 1950; the figure had probably doubled by 1972, although data are not yet available.

Meat Packing Industry. In meat packing, modern slaughtering and meat processing practices reduce the waste load, and more packing plants are expected to use biological and chemical waste treatment in addition to conventional catch basins, which only have provisions of gravity settling and grease skimming [24]. In 1950, an estimated 20 percent of all packing plants had no waste treatment facilities; the percentage without facilities is expected to drop to one percent by 1977. In 1967, 70 percent of the 75.3 billion gallons of waste water emerging from packing plants was routed to municipal treatment facilities; by 1977, 85 percent of meat packing waste water is expected to receive this treatment. In poultry processing, which includes the slaughtering, dressing, and packing of chickens, broilers, and turkeys, waste discharges are decreasing due to more efficient processing methods, increased use of byproducts, greater utilization of on-site biological treatment of wastes, and wider use of municipal treatment facilities (anticipated for 80 percent of poultry waste water by 1977 as opposed to 50 percent in 1950).

Point of Decision. As exemplified by the dairy and meat packing industries, however, more efficient, cleaner processing operations have been prompted as much by financial considerations as by concern for environmental side effects. The result is a relative improvement in environmental conditions, but only because economic and environmental considerations happen to be parallel. This beneficial concurrence cannot be relied upon in all industries. As we have seen in the steel and canning industries, new technology introduced on the basis of economic priorities can be more polluting than older production methods. The point is that if there is to be a true balance between industrial and environmental demands, then an analysis of the costs of continued pollution versus the benefits of the technology causing that pollution must play an integral part in the decision processes which result in the introduction of new technology, alteration of the old, and selection of alternatives. Otherwise, we may ever be frantically looking for new waste control methods to deal with already existing pollution. The waste that emerges may resist conventional treatment methods since the original product was created with marketing potential in mind rather than with an eye to environmental demands. An example is the creation of synthetic organic chemicals which, as observed earlier, sometimes cannot even be identified by their creators and which pass intact through conventional waste treatment facilities. The lack of consideration for the environmental impact of manufacturing processes and the related failure to establish tailor-made pollution control devices for specific products has forced us to fall back to a large degree on broad-scale, conventional waste treatment methods such as land disposal, deep-well injection, incineration, and biological treatment. Each of these methods, in turn, produces environmental problems of its own in the form of further water or air pollution.

These difficulties need to be kept in mind since they are not reflected in the otherwise encouraging trends anticipated in organic wastes discharged to waterways from the five industries just discussed. These trends, in millions of pounds after industrial or municipal treatment, are shown on the next page in the comparison between high-organic waste discharges from these industries to waterways in 1963 and as projected for 1977 [25].

Pulp and Paper Industry. One major polluting enterprise quite well advanced in the development of pollution controls is the pulp and paper industry. Application of the controls has varied, however. A study published in

		BOD	Suspended Solids	Total Dissolved Solids
Dairy Products	1963	931.8		
	1977	61.52		
Cotton Finishing	1963	528.0	166.0	498.0
	1977	516.0	160.0	494.0
Wool Weaving and Finishing	1963	69.4		
	1977	52.9		
Meat Packing	1963	454.0		
	1977	147.0		
Poultry Processing	1963	17.2		
	1977	10.2		

1971 by the Council on Economic Priorities assessed treatment applied to pollution from 131 pulp mills producing some 87,000 tons of pulp per day, equal to approximately 68 percent of the total daily 1969 industry production [26]. The 131 mills used at least 2.4 billion gallons of water daily, yet 833.5 million gallons of waste water each day—or about 35 percent of the total daily discharge—received inadequate primary treatment or no treatment at all. The mills removed heavy solids from another 29 percent of the effluent, ignoring the BOD potential and chemical pollutants. On the other hand, a more recent study by the Council on Economic Priorities lauds the pulp and paper industry for making considerable progress in the past few years under the pressures of public concern and stiffer regulations. The council estimates that, by 1975, most pulp mills will have up-to-date pollution control equipment which, in many cases, will consist of secondary treatment [27].

Lack of data, however, makes it difficult to assess the whole industry, which includes the pulp milling process in which wood is prepared, pulped, screened, washed, thickened, and bleached, and the paper mill function in which the stock is prepared and the paper is made and finished. These processes produce large amounts of BOD wastes, solid wood residue, sap from washed wood, and chemicals. In 1967, only an estimated ten to 15 percent of pulp and paper industry effluent discharge went to municipal sewers; water pollution control then relied mainly on in-plant water reuse, chemical and solid waste recovery, and processing improvements [28]. Increasing use is now being made of on-site primary and secondary treatment to reduce BOD and suspended solid waste loads, but tertiary treatment to remove chemicals and inorganic nutrients is used in only a handful of companies. The need for decisive control of pollution from this critical industry is clear from the projected gross waste loads expected in coming years. Newer technology in the industry is less polluting, but this advantage is expected to be offset by in-

One of the producers of the millions of tons of paper for printing and packaging, the pulp mill ranks among the worst polluters of air and water. In the past few years, under the pressure of public concern and stiffer regulations, progress has been made in reducing pulp mill effluents.

creased paper production, resulting in an overall rise in the amount of wastes produced during production and consequently putting a greater burden on waste treatment measures. Gross waste loads in the U.S. pulp and paper industry are shown below for 1968, 1972, and as projected for 1977 (based on wastes produced per ton of product *before* waste treatment) [29].

Gross Waste	Suspended Solids*	Dissolved Solids*	Total Solids*	BOD*	Wastewater Quantities (billions of gallons)
1968	2.82	11.1	13.8	2.71	930
1972	3.03	11.8	14.8	2.91	990
1977	3.50	12.9	16.3	3.21	1050

*In billions of pounds.

Agriculture and Runoff. The water pollution impact of another agricultural activity—cattle and poultry raising—also is unclear. The potential is serious, since farm animals produce ten times as much waste as the human population in the United States [30]. Modern feeding methods that emphasize highly concentrated feed and small confinement areas for some phases of the animal's feeding cycle intensify the problem; in range feeding, wastes are more readily dispersed and converted into soil nutrients by natural processes. Organic animal wastes carry human disease organisms such as salmonella and also deplete dissolved oxygen in streams. Under current practice, runoff from animal wastes generally is allowed to flow directly into waterways, or is circulated through anaerobic basins where some biological degradation takes place, or is spread in pastures or elsewhere on land. Well dispersed land application is the most effective waste disposal measure, reducing pollution entering streams by runoff by as much as 99 percent. Anaerobic basins are generally unsatisfactory since they produce an effluent that exceeds raw domestic sewage in BOD strength. The direct discharge of animal wastes into streams, practiced in some areas with dairy and swine herds, causes severe pollution. The discharge may result from the intentional channeling of wastes from feeding areas into the streams or from roving animals given access to the watercourses. The amount of contamination eventually reaching a waterway thus depends upon a number of farm management factors, as well as on soil type and land contours, so that generalizations are difficult without a more comprehensive picture of farming practices. A field study in North Carolina, however, is useful in assessing the relative amounts of various pollutants that can be expected from different animals according to different waste management practices. The table that follows, taken from this study [31], shows average amounts of wastes in runoff from representative North Carolina animal growing operations.

Operation	Fecal Contamination*	BOD†	Total Organic Carbon**	Nitrogen†	Phosphates†
Swine					
Direct Discharge (1)	30,000.0	.176	.128	.032	.017
Anaerobic Lagoon (2)	910.0	.154	.156	.009	.005
Land Spreading (3)	630.0	.003	.007	.002	.001
Dairy (1)	60,000.0	.190	.230	.034	.027
Beef (3)	430.0	.030	.100	.008	.003
Poultry (3)	0.6	.0005	.0007	.0002	.0003

(1) Sampling station downstream from point of discharge, with intervening section of stream providing some stabilization and settling.
(2) Basin (lagoon) effluent.
(3) Sampling station at base of watershed immediately below site.

* Millions of coliform bacteria per animal per day.
** In pounds per animal per day.
† As a measure of oxygen-demanding wastes in pounds per animal per day.

The modern method of feeding cattle, emphasizing highly concentrated feed and small confinement areas, intensifies water pollution, while the manure that used to return to the land must now be replaced by chemical fertilizers. At the next stage in moving the meat to market—the packing houses—another water pollution problem is created.

In addition to increasing the direct discharge of animal wastes, current agricultural practices tend to pollute more than older farming methods because today farmers rely more heavily on chemical fertilizers and pesticides (both of which are discussed elsewhere). At this point, it is sufficient to add agriculture to our tally of the major human activities in which new technology goes hand in hand with intensified water pollution.

Pressures for Change in Water Treatment

Generally, this intensification has steadily increased the pressure on industrial and municipal waste treatment systems, since more fundamental changes to reduce the pollution capacity of modern technology have been so limited. In some cases such as steelmaking, canning, chemical manufacturing, and farming, the changes that have been made actually have increased the pollution load or toxicity. Reliance on waste treatment systems to purify the effluent, however, has been disappointing for two reasons: (1) the science of waste treatment has lagged far behind the capacity to produce complex and voluminous industrial pollutants, and (2) available control devices, imperfect as they may be, are very expensive to install and operate.

Conventional industrial waste treatment equipment tends to borrow heavily from sanitary waste treatment methods such as mechanical screening, settling, and skimming common to primary treatment; biological treatment employing trickling filters or activated sludge to reduce BOD, as is done in secondary treatment; and application of various chemicals such as alum or lime to remove small waste particles by coagulation and subsequent sedimentation, as is done in municipal tertiary treatment for inorganic phosphates. Industrial organic matter that resists biological degradation can sometimes be reduced by *adsorption* (adherence due to surface attraction) on granulated or powdered carbon, an approach also used in advanced municipal waste treatment and in drinking water purification. Methods to separate liquid materials of different weights by gravity or by centrifuge to obtain the desired product and isolate the waste have their origin in commonly applied scientific principles. Other forms of advanced waste treatment include the use of special resins to remove inorganic ions, electrodialysis to remove salts, advanced mechanical filters in combination with water soluble compounds to control the consistency of the material being filtered, and a process called ammonia stripping which allows nitrogen-containing ammonia gas to escape from water into the atmosphere and thus reduces the amount of nitrogen in wastes that otherwise might be incorporated in nitrates—inorganic nutrients that also may be a health threat in drinking water. These processes are used in both municipal and industrial waste treatment, but generally have been applied most often in industry.

Some waste treatment processes have been tailor-made for specific industries: the removal of sulfides in tannery wastes by the use of flue gases containing carbon dioxide; the complexing of cyanides with ferrous compounds to render the cyanides insoluble; extracting phenols with solvents; distilling ammoniacal liquors; the removal of salts by forcing water to reverse its normal osmotic passage (reverse osmosis) through membranes that are not permeable to dissolved salts; and precipitating metals with chemicals. These techniques are generally expensive, however, and industry, by and large, has relied instead upon the broad-scale treatment methods developed to reduce the BOD load in domestic sewage. Thus, despite sharply rising investments in waste control equipment, companies frequently have been unable even to hold the line in waste containment. As we have seen, this situation has prompted more and more industries to pump their waste streams into the municipal treatment system, passing on the problem to the community as a whole.

This practice has inspired a heated controversy. The argument that heavy industrial use of municipal facilities represents a public subsidy for industry is countered by the argument that such centralized treatment is more efficient than onsite treatment which entails higher costs that would be passed on to the public anyway in the form of higher product prices. Furthermore, the proponents argue, user charges paid by industry in some areas on the basis of plant-waste characteristics are more than adequate to cover operating costs resulting from the discharges.

A major complaint on the part of some sanitary engineers is that the toxic substances contained in industrial wastes present a special environmental threat and may even disrupt the municipal treatment process. For example, a change in the composition of paper mill plant wastes resulted in the buildup of partially unstabilized organic material in the sludge drying beds of a Kalamazoo, Michigan, municipal plant in 1967 [32]. This produced a putrid odor that required costly countermeasures which reduced the effectiveness of the plant. A serious fire broke out in one Ohio plant, presumably on ignition of volatile industrial wastes [33]. In other plants, poisonous chemicals and metallic compounds from industry have interfered with bacteria that decompose wastes.

Other sanitary engineers argue that such episodes, although occasionally spectacular, have been limited in scope and will become increasingly rare as stricter municipal codes are enacted to prevent or limit sewer dischage of toxic substances, oils, and gases; the codes will force industries to pre-treat wastes that contain such materials.

Use of Public Money

Interwoven with these problems is the more subtle and still unresolved issue about the extent to which federal funds, which can be granted for 75 percent of the total cost of treatment facilities, should be used for industrial waste treatment capacity. Studies by the U.S. Department of the Interior and the U.S. General Accounting Office (GAO), a Congressional watchdog agency, have made it clear that a considerable amount of federal money has been used in this way [34]. In one study, 44 of 381 facilities were found to be designed to treat wastes that were at least 75 percent of industrial origin. Total estimated cost of the 44 facilities was $46.3 million, with various federal grants contributing about $17.2 million. In one case, grants totaling $806,000 were given to build a waste treatment plant for the sole use of a food-processing plant, apparently as an inducement for the plant to locate in one particular area. The treatment plant was built, but, ironically, had to be put out of service in 1969 when the food-processing plant halted operations. In its own survey, the GAO identified seven projects, out of 44 audited, which were for the construction of municipal facilities to treat exclusively industrial wastes. Federal construction grants for the seven projects totaled about $503,000. For 29 of the remaining 37 projects, industrial wastes comprised 50 percent or more of either the total volume of waste or of the BOD removal capacity of the plants—a percentage equivalent to the estimated national average of industrial waste loading in municipal waste treatment plants. Proponents admit that occasional questionable applications of funds for industrial waste treatment have been made in the guise of municipal facilities, but they quickly add that the distinction between municipal and domestic wastes is artificial. For example, commercial and service industries such as laundries, restaurants, and hotels have customarily been included as legitimate sources of municipal wastes. The line between these commercial interests and those represented by

New York City's $228 million North River Pollution Control plant at 135th Street on the Hudson River is scheduled for completion in 1979. Large and small municipalities are investing more and more funds in water treatment plants, accompanied by much debate over the extent to which these plants should bear the burden and the cost of treating industrial waste.

groceries, department stores, warehouses, and industrial parks seems difficult and arbitrary, the proponents argue.

Once past these generalities, the debate over the industrial use of municipal treatment facilities involves complex legal, economic, and social questions beyond the scope of our discussion. The issue is relevant here because it illustrates the degree to which municipal waste treatment plants have become the focus of water pollution control efforts. Stricter water quality standards, increased industrial loadings, population growth, and inflation have been cited as reasons for the ever rising level of investment in municipal treatment plants. Projected investment needs indisputably have increased, although the reason for this is unclear. In 1960, an estimated $2 billion was needed to overcome a so-called "backlog" of required treatment facilities in the United States, according to a survey of the Conference of State Sanitary Engineers. The backlog has been one of the most enduring concepts in water pollution control strategy and its resolution has been one of the most elusive goals of control strategists [35]. The backlog concept influenced estimates of the proposed investment level under the amended Federal Water Pollution Control Act during the mid-1960s, but by 1967 the FWPCA had already raised the estimated total investment need to $8 billion to provide secondary waste treatment for most of the urban population of the nation.

The backlog gap has steadily widened since. In 1970, the federal government established a new goal of secondary waste treatment where needed and advanced treatment as necessary in special cases. The price tag for the five years from 1970 to 1975 was set at $10 billion. By the end of 1970, however, the states already were requesting a total of $12.6 billion by 1975 to meet upgraded water quality standards [36].

By the end of 1971, a total of some $8.2 billion in federal, state, and local grants for municipal treatment facilities had been approved in the 11 years since the original backlog need of $2 billion was identified by the Conference of State Sanitary Engineers survey [37]. Despite the fact that more than four times this original estimate had already been authorized for treatment facilities, state representatives estimated in 1972 that $18.1 billion would be needed from mid-1971 to mid-1976 to meet new community construction plans designed to fulfill anticipated 1976 water quality standards [38].

By comparison, $3.0–7.1 billion in private capital was estimated to be needed for industrial pollution control installations as of 1971. Information about industrial investment for previous years is sketchy, but one EPA estimate indicates that major manufacturers (excluding utilities and mining) spent roughly $2.2 billion on water pollution controls during 1967–1970 [39], compared with $4.2 billion in public funds earmarked for municipal treatment plants [40].

Need for Higher Quality Water Standards

At any rate, the expanded concept of centralized municipal plants to treat a range of community needs, including those which industry cannot or will not meet, is one factor which explains why by 1976 we will have invested perhaps $26.3 billion for treatment plants instead of the 1960 estimate of $2 billion. Even a four percent increase per year due to inflation would boost the 1960 estimate to only $3.6 billion by 1975, still sevenfold than the projected 1976 figure. Allowing for a broadened sense of community service by the plants, this discrepancy still seems to indicate a serious miscalculation in 1960—a miscalculation that many communities attribute to the need for higher water quality standards. Contributing to this, however, would be the outdated concepts of water pollution potential that existed in 1960, as well as inadequate pollution regulations.

Further evidence of a serious miscalculation was the tacit admission in 1970, in face of increased public concern, that even secondary treatment will not be adequate everywhere since some regions will need advanced treatment. Advanced treatment facilities generally can be added to existing secondary treatment plants, but only at considerable cost. For example, the plant at Lake Tahoe has been equipped with one of the most comprehensive treatment systems in existence to protect the purity of that famous natural lake [41]. The 1969 capital replacement costs (calculated to provide a common basis for comparison) for the primary, secondary, sludge disposal, and chlorine disinfection phases of the system were $2.6 million. By comparison, capital replacement costs for the advanced processes in the treatment facility were $2.9 million. (These included lime treatment, ammonia stripping, filtration, carbon adsorption, and methods to recondition used lime and carbon.)

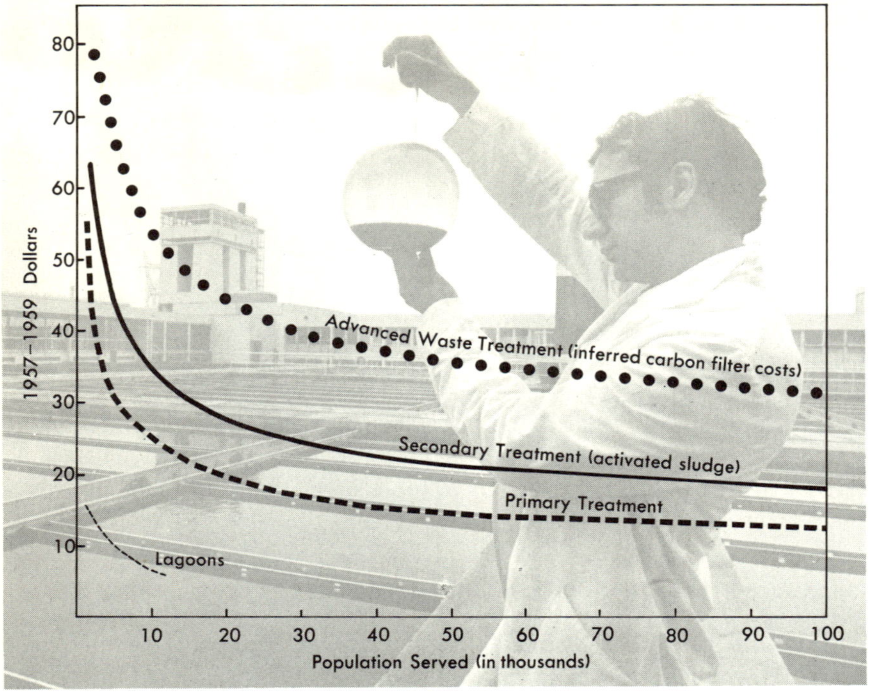

Generalized per capita construction cost of primary, secondary, and advanced waste treatment processes.

From February 1969 to January 1970, operating, labor, and repair costs for the phase of the system comparable to conventional municipal treatment were $103.05 per million gallons of daily waste flow into the plant; for the advanced treatment phase the comparable figure was $136.51. These costs cannot be automatically applied to other plants, however, because the Tahoe costs tend to be higher than elsewhere; the plant itself is small, which actually increases overall operating costs in comparison with larger plants that benefit from better purchase rates for larger quantities of materials. The relative increase in capital and operating costs produced by the introduction of advanced waste treatment, however, demonstrates the degree to which investment needs will again rise if advanced treatment becomes a nationwide goal. On the national level, the federal estimate in 1957–1959 dollars shown in the following graph demonstrates the sharp increases in construction costs that can occur if just one advanced waste treatment method—carbon filtration—is added to the process. The estimate is probably conservative due to the effect of subsequent inflation, and it does not, of course, take into account the other advanced techniques that supplement carbon filtration at Lake Tahoe.

In frustration over the mounting backlog burden, officials of the EPA's

Water Quality Office wrote in a 1971 report: "An immutable tendency seems to be that as federal financial assistance and investment increase, physical plant expands; and as physical plant expands, the volume of capital needs involving federal financial assistance also expands. The more we invest, the more we seem to need to invest." [42]

This feeling was substantiated by the discovery of an apparent trend toward overbuilding in municipal plants, particularly the expansion of capacity at already existing facilities. The EPA officials observed there is room for disquiet:

> There is room for disquiet when one takes into account the fact that fully a quarter of metropolitan area waste treatment plant capacity is less than half utilized, and that for nonmetropolitan communities, over 30 percent of total waste treatment plant capacity is utilized at less than half of design rating. When one excludes the one-sixth of all waste treatment plants that are overloaded, the mean utilization rate for publicly operated plants in the United States is found to be just under 63 percent—almost two-fifths of the total capacity of plants of every vintage, then, is simply unused. Worse, in terms of aggregated probabilities, much of it will never be used. The formal useful life of a waste treatment plant is 25 years. At the rate of population growth that applied during the 1950s [which has steadily declined since] only the fastest growing classes of communities could make full use of the capacity of a plant designed to serve twice its initial loading. [43]

These misgivings seemed particularly acute to the EPA analysts in view of the need for financial resources for other pollution control measures. Our own purpose is not to judge the validity of the municipal plant program, but to identify the elements that must be weighed in any analysis that seeks balanced allocation of pollution control resources. Such an analysis simply has not been made in the case of the treatment plant program, primarily because sanitary waste treatment at centralized facilities in the past has always been considered the mechanism to remove pollution, and the emphasis has changed very little. This is understandable in view of the history of water pollution control, but, although nurtured by many intelligent, dedicated professionals, the pollution control concepts of the 1960s clearly have failed to provide a degree of control acceptable today.

A related area of investment also has not had a thorough hearing in terms of cost-benefit analysis. That is the emphasis on the separation of old combined sanitary-storm sewers in which street runoff and domestic wastes are mixed together to produce rates of flow during heavy rains that are too large to be treated at municipal plants and thus must be diverted directly into waterways untreated. In 1967, it was estimated that correcting this combined sewer overflow problem by separating sanitary and storm sewers would cost more than $47 billion [44]. This amount, of course, would be in addition to the $18.1 billion which may be needed for municipal treatment plants at the end of the sewers. Compromise solutions such as high-rate treatment processes for combined flows and overflow regulators have since been suggested to reduce this cost, but considerable opinion still favors the physical separation of waste streams, with untreated storm drainage piped into the rivers. The awkward part of this method is that storm waters carry a great deal of pollution which, if discharged untreated, as suggested, can counteract the improvements in stream quality that the advanced treatment plants at the end of the sewage stream are designed to produce.

Urban pollutants which enter the storm waters include street litter, salt and other de-icing chemicals, pesticides, fertilizers, feces and urine from pets, household and commercial refuse, and waste waters from air-scrubbing operations that convert air pollutants into water pollutants. One survey in Chicago indicates that soluble dust and dirt in the streets of a large city contain a BOD potential equivalent to five percent of that in the secondary treatment effluent from the municipal plant serving people in the city [45]. Much of the dirt and dust is from human activities such as construction, tire shredding, and road surfacing material, rather than from natural sources. In less densely populated urban regions, the BOD load of storm sewer water actually may equal the BOD load in the local secondary wastewater effluent [46]. Lead and hydrocarbon wastes from vehicle exhaust and other air pollution fallout also are prominent contaminants of storm water [47]. The contribution from pets should not be underestimated. Some one-half million dogs in New York City leave 5000–20,000 tons of feces and 600,000–1 million gallons of urine in the city streets each year. In Baltimore, the figures for 100,000 dogs are 4000–13,000 tons of feces and 120,000–200,000 gallons of urine per year [48]. These organic wastes present some human health hazard as well as add a significant BOD load to waste water.

Alternative Waste Treatment Methods

Sanitary waste treatment is, of course, essential to human health. Lacking, however, has been a thorough study of alternative methods too achieve sanitary treatment with the fewest possible environmental side effects. One exploration of comprehensive systems for sanitary waste treatment was recently completed by the U.S. Army Corps of Engineers [49]. Three different treatment systems and appropriate sewer networks were assessed for the 2500 square-mile Chicago-Northwest Indiana region which has 13,500 industries and will have an estimated nine million persons by 1990. Each alternative system was investigated at a different level of application, but the most complete would involve, by 1990, treatment of an average of some 3.3 billion gallons per day of municipal, industrial, stormwater, and agricultural wastes. This total, incidentally, is based on the assumption that industries will greatly reduce water use by 1990 through extensive in-plant water recycling.

One alternative studied was advanced biological treatment as an adjunct to secondary treatment facilities already in existence. A second alternative, specialized physical-chemical techniques, required the replacement of all existing facilities. The third was a land treatment plan that called for the transport of wastes via a tunnel system to a 700-square mile site some 25 miles south of Chicago. There, the raw wastes would be treated biologically, then stored in lagoons, and later applied to land at controlled rates to supply the critical nutrient requirements of food crops during the growing season. Irrigation water purified by percolation through the soil would be collected in a drainage system and then piped back to the Calumet River drainage system near Lake Michigan to recharge the water system from which it was originally drawn for domestic and industrial use. Sludge generated by biological treatment of the wastes would be used, after microbial digestion, as nutrients and soil builders.

Cost estimates of the three alternative systems were made, using various construction bid data, experience with pilot waste disposal projects, and costs at the Muskegon County land disposal wastewater management project [50]. As seen in the table below [51], costs for comprehensive land treatment by irrigation were by far the lowest, although the price tag for each system was staggering. The $7.5 billion capital cost for the advanced biological treatment system, for example, was exactly one-third of the highest amount estimated by the government to be needed during 1970–1974 for municipal and industrial sewers and treatment facilities *throughout the entire country* [52]—another indication of the seemingly limitless investment that can be made in wastewater treatment systems.

	Advanced Biological Treatment	Physical-Chemical Treatment	Land Treatment
Capital Costs*	$7,480,000,000	$6,745,000,000	$5,357,000,000
Annual Costs** (debt service, operation and maintenance, major replacements)	729,000,000	673,000,000	424,000,000
Annual Cost per Capita (on the basis of nine million population in 1990)	81.00	74.80	47.10

The land treatment system also produces an effluent (actually, the drainage water from irrigation) that, in comparison with the effluent from the other two treatment methods, is lower in BOD, suspended solids, soluble phosphorus, ammonia-nitrogen, oils, greases, phenols, disease organisms, and heavy metals, according to performance experience in small-scale operating systems and in pilot plants. Another advantage to the land system would be the reduction in the amount of air pollution it produced in comparison with the other methods, particularly the physical-chemical system that each day would discharge from process stacks up to 340 tons of nitrous oxides, ten tons

*In 1972 dollars.

**Estimated for the period 1975–2024, allowing for a rise in yearly expenses; each system designed to treat municipal, industrial, stormwater, and rural agricultural runoff at flow rate expected in 1990.

of sulfur dioxide, and 200 tons of particulates. Equipment such as pumps and dredges at the land treatment site would produce fewer pollutants, although the exact amount was not determined in the Army Corps study.

Crop irrigation with domestic wastes is attractive from an ecological point of view since it returns nutrients to nature in a constructive way. Theoretically, this method comes much closer to completing the natural cycle than disposing of wastes in stream water, since food nutrients originally obtained by man in the United States most often come directly from the land in the form of crops or indirectly in the form of grain-fed animals. Although the land treatment theory is sound, there are problems in its implementation. First of all, soil has only a limited purification capacity that is dependent upon a number of factors, probably the most important being the amount of oxygen present to support the natural microbial decomposition of wastes [53]. A modern problem is that municipal wastes—particulary sludges—often contain industrial contaminants, among them heavy metals such as zinc and copper which can build up in the soil to concentrations that are poisonous to plants [54]. In addition, there is some evidence that cadmium and lead contaminants in the wastes could be taken up by plants in amounts that might prove hazardous to man or animals [55].

The irrigation alternative, while clearly worthy of considerable study [56], once again demonstrates that nature will accommodate man only to a certain degree. That is, the natural biological cycles are beneficial to us only as long as we do not disrupt them with large quantities of poisonous inventions from factory and laboratory which cannot be changed into substances that support life.

Faulty Control Strategy

Historically, municipal treatment plants have been regarded as places to pipe human wastes, and this has lured us into heavily emphasizing those facilities in overall pollution control strategy without much critical thought. The very term "pollution control" implies that the basic solution lies in control and not in dealing with the original sources and causes of pollution. Despite their own commitment to the municipal treatment program, EPA officials mused in one of their 1961 reports:

> The tentative conclusion that waste treatment is no more than a convenient point of departure for any meaningful strategy of water pollution control is reinforced by examination of the sources of recent increases of pollutants. Underlying the growth of available biochemical oxygen demand and of phosphorus are basic economic forces. To counter the polluting effects of fundamental features of the twentieth century, technology and social organization would seem to call for fundamental remedies. [57]

The examples of polluting activities here and in the preceding chapter only intimate the strong economic and social forces generating those activities. There is room here only to point out the clear need for a better under-

Water pollution control has long focused on the output end of the sewer pipe, but we will not solve our water problems without a more serious examination of—and attention to controlling—what enters the input end.

standing of those forces in plotting our strategy to clean up the waters of the world. Just as many municipalities are bedeviled by the unexpected difficulties caused by leaking sewer pipes, our control strategy has also sprung many unanticipated leaks. Thus, controls on marine oil pollution neglect huge influxes of hydrocarbons from air pollution. Crackdowns on chemical discharges into waterways ignore the large amounts dumped legally at sea as waste. Attempts by port authorities to achieve some control over the waste cargoes of disposal barges headed out to ocean dumping grounds are counteracted by pipes not far away that discharge raw sewage and industrial wastes in near-shore waters. Efforts to update sewage treatment plants to disinfect human wastes overlook large contributions from domestic animals using the same streams. Heavy investment in some industries for pollution controls are counteracted by more polluting technology in the same industries. Improved techniques to treat organic wastes with a heavy BOD load are not designed to remove organic chemicals that are serious biological threats. Opening the sewer pipes to industrial wastes in order to insure more uniform treatment also opens the gate to disruption of the biological treament process by industrial poisons and unlocks the door to vested interest pressure from businesses that stand to benefit from expensive new treatment facilities paid for by the public. Separation of sanitary and storm sewers to achieve better domestic waste purification neglects the pollution load of storm water. Finally, large expenditures to insure excellence in just one area of pollution control—municipal treatment—as necessary as that treatment is, encourage spending of ever larger amounts of money to expand the capacity of this one method in the hopes that it will prove a cure-all.

All of this effort neglects a more critical look at the input end of our symbolic sewer pipe where the fundamental processes and living habits of man create the waste problem in the first place—a problem that clearly has outgrown anything but the most gigantic treatment schemes. Analysis of that root problem is not our purpose here, but the importance of such an analysis is emphasized by this material. As the final irony, when we reach the other end of the pipe where the secondary treatment municipal plant is located, we discover that even this control has a major leak. This one promotes the widespread aquatic illness called eutrophication, which will be the subject of the next chapter.

[1] Barry Commoner, *The Closing Circle* (New York: Alfred A. Knopf, 1971).

[2] R. Norris Shreve, *The Chemical Process Industries,* Second Edition (New York: McGraw-Hill, 1956), p. 6.

[3] Nathaniel Wollman and Golbert W. Bonem, *The Outlook for Water: Quality, Quantity, and National Growth* (Baltimore: Johns Hopkins Press, 1971), adapted from Table 18, p. 58. (Values are based on the "medium" level of predicted population and economic growth, as opposed to "low" or "high" estimates.)

[4] *The Cost of Clean Water, Volume II: Cost Effectiveness and Clean Water,* Environmental Protection Agency (Washington, D.C.: U.S. Government Printing Office, March 1971), pp. 1–128. The pollution summary by the EPA has been assessed by the Council on Environmental Quality (CEQ) as having advantages in determining the situation throughout the United States, but as being flawed by inconsistencies in the use of raw data and by vagueness about the specific types of pollutants involved. The deficiencies have been corrected to a certain extent in the 1972 EPA report (see [38]), but changes in methodology in the 1972 report make it more difficult to use for comparative purposes in this chapter. The CEQ criticism, and another less comprehensive pollution index commissioned by CEQ, can be found in *Environmental Quality, The Third Annual Report of the Council on Environmental Quality* (Washington, D.C.: U.S. Government Printing Office, August 1972), pp. 11–16. Judging by the quality of current information, adequate overall data on extent and trends of water pollution throughout the United States will not be available for several years.

[5] *The Cost of Clean Water and Its Economic Impact,* Volume 1, U.S. Department of the Interior (Washington, D.C.: U.S. Government Printing Office, January 10, 1969), pp. 1–220.

[6] *Acid Mine Drainage in Appalachia,* report of the Appalachian Regional Commission, 1666 Connecticut Avenue, Washington, D.C. 20235, 1969, p. *iii.*

[7] *Cost of Clean Water: Cost Effectiveness and Clean Water,* p. 38 (see [4]).

[8] Ibid., p. 30.

[9] *The Cost of Clean Water, Volume III: Industrial Waste Profiles No. 1—Blast Furnace and Steel Mills,* FWPCA Publication No. I.W.P.-1 (Washington, D.C.: U.S. Government Printing Office, September 1967), pp. 1–100. Table prepared from data on pp. 55–56.

[10] A discussion of the respective roles of population, affluence, and technology in the increase of pollution can be found in *The Closing Circle,* especially in Chapters 8 and 9, pp. 125–77 (see [1]).

[11] *Cost of Clean Water: Blast Furnace and Steel Mills,* table prepared from data on pp. 79 and 82 (see [9]).

[12] *The Cost of Clean Water, Volume III: Industrial Waste Profiles No. 6—Canned and Frozen Fruits and Vegetables,* FWPCA Publication No. I.W.P.-6 (Washington, D.C.: U.S. Government Printing Office, September 1967), pp. 1–104.

[13] Julian McCaull, "Know Your Enemy," *Environment,* June 1971, Vol. 13, No. 5, 30–39.

[14] *Cost of Clean Water: Canned and Frozen Fruits and Vegetables,* prepared from data on pp. 53 and 88 (see [12]).

[15] Arthur D. Little, Inc., *Organic Chemical Pollution of Fresh Water, Water Quality Criteria Data Book,* Volume 1, prepared for the Environmental Protection Agency (Washington, D.C.: U.S. Government Printing Office, December 1970), pp. 1–379.

[16] Lawrence H. Keith, "Chemical Characterization of Industrial Effluents," paper presented at the annual American Chemical Society, Division of Water, Air, and Waste Chemistry, Boston, Massachusetts, April 1972.

[17] Ibid.

[18] Neil A. Iliff, "Organic Chemicals (Except Agricultural, Animal Health Chemicals, and Food Additives)," paper presented at the 2nd International Symposium on Chemical and Toxicological Aspects of Environmental Quality, Munich, May 27–28, 1971.

[19] *The Cost of Clean Water, Volume III: Industrial Waste Profiles No. 7—Leather Tanning and Finishing,* FWPCA Publication No. I.W.P.-7 (Washington, D.C.: U.S. Government Printing Office, September 1967), pp. 1–60.

[20] Ibid., p. 26.

[21] Ibid., from Table 3, p. 37.

[22] *The Cost of Clean Water, Volume III: Industrial Waste Profiles No. 9—Dairies,* FWPCA Publication No. I.W.P.-9 (Washington, D.C.: U.S. Government Printing Office, September 1967), pp. 1–166.

[23] *The Cost of Clean Water, Volume III: Industrial Waste Profiles No. 4—Textile Mill Products,* FWPCA Publication No. I.W.P.-4 (Washington, D.C.: U.S. Government Printing Office, September 1967), pp. 1–133.

[24] *The Cost of Clean Water, Volume III: Industrial Waste Profiles No. 8—Meat Products,* FWPCA Publication No. I.W.P.–8 (Washington, D.C.: U.S. Government Printing Office, September 1967), pp. 1–118.

[25] *Cost of Clean Water, Volume III,* p. 164 (see [22]); pp. 26 and 69 (see [23]); pp. 40 and 50 (see [24]).

[26] "Paper Profits: Pollution in the Pulp and Paper Industry," *Economic Priorities Report,* Vol. 1, No. 6, Council on Economic Priorities, Washington, D.C. December 1970–January 1971, 47 pages.

[27] "Paper Profits: Pollution Audit 1972," *Economic Priorities Report,* Vol. 3, No. 3, the Council on Economic Priorities, Washington, D.C. July–August 1972, 51 pages.

[28] *The Cost of Clean Water, Volume III: Industrial Waste Profiles No. 3—Paper Mills,* FWPCA Publication No. I.W.P.–3 (Washington, D.C.: U.S. Government Printing Office, November 1967), pp. 1–65 and appendices.

[29] Ibid., adapted from Table 5, p. 28.

[30] *Role of Animal Waste in Agricultural Land Runoff,* prepared for the Environmental Protection Agency by the Department of Biological and Agricultural Engineering, School of Agriculture and Life Sciences, North Carolina State University, Raleigh (Washington, D.C.: U.S. Government Printing Office, August 1971), pp. 1–114.

[31] Ibid., from data on pages 34 and 94.

[32] *Cost of Clean Water and Its Economic Impact,* p. 94 (see [5]).

[33] *The Economics of Clean Water,* Volume 1, Environmental Protection Agency (Washington, D.C.: U.S. Government Printing Office, 1970), p. 138.

[34] *Federal Grants Awarded for Constructing Waste Treatment Facilities which Benefit Industrial Users,* Comptroller General of the United States, Washington, D.C. 20548, May 8, 1970, pp. 1–34.

[35] *Cost of Clean Water and Its Economic Impact,* p. 21 (see [5]).

[36] *Cost of Clean Water: Cost Effectiveness and Clean Water,* p. 4 (see [4]).

[37] *Water Pollution Control Programs. Hearings before the Subcommittee on Air and Water Pollution of the Committee on Public Works* (Washington, D.C.: U.S. Government Printing Office, February 4, 8, and 9, 1971), pp. 464–65.

[38] *The Economics of Clean Water: Summary of Analysis,* Environmental Protection Agency (Washington, D.C.: U.S. Government Printing Office, 1972), pp. 1–33.

[39] *Cost of Clean Water: Cost Effectiveness and Clean Water,* pp. 7–9 (see [4]).

[40] *Water Pollution Control Programs,* pp. 464–65 (see [37]).

[41] Russell L. Culp and Gordon L. Culp, *Advanced Wastewater Treatment* (New York: Van Nostrand-Reinhold, 1971), pp. 1–310.

[42] *Cost of Clean Water: Cost Effectiveness and Clean Water,* p. 1 (see [4]).

[43] Ibid., p. 72.

[44] *Water Pollution Control Programs,* p. 377 (see [37]).

[45] American Public Works Association, *Water Pollution Aspects of Urban Runoff,* prepared for the Federal Water Pollution Control Administration (Washington, D.C.: U.S. Government Printing Office, January 1969), p. 3.

[46] Edward H. Bryan, "Quality of Stormwater Drainage from Urban Land," paper presented at the 7th American Water Resources Association Meeting, Washington, D.C., October 28, 1971.

[47] G. Söderlund, H. Lehtinen, and S. Friberg, "Physicochemical and Microbiological Properties of Urban Stormwater Runoff," paper presented at the 5th International Water Pollution Research Conference, San Francisco, July-August 1970.

[48] Alan M. Beck, School of Hygiene and Public Health, The Johns Hopkins University, Baltimore, Maryland, personal communication.

[49] *Regional Wastewater Management Systems for the Chicago Metropolitan Area, Summary Report and Technical Appendix,* Office of the Chief of Engineers, Department of the Army, Washington, D.C. 20314, March 17, 1972.

[50] For example, see *The Muskegon County Wastewater Management System,* Muskegon County Department of Public Works, Muskegon, Michigan, pp. 1–16.

[51] *Regional Wastewater Management Systems for the Chicago Metropolitan Area* (see [49]).

[52] *Controlling Industrial Water Pollution—Progress and Problems,* Comptroller General of the United States, Washington, D.C. 20548, December 2, 1970, p. 12.

[53] Karl Imhoff, W. J. Müller, and D. K. B. Thistlethwayte, *Disposal of Sewage and Other Waterborne Wastes,* Second Edition (Ann Arbor, Michigan: Ann Arbor Science Publishers, Inc., 1971), p. 104.

[54] T. D. Hinesly, O. C. Braids, and J. E. Molina, *Agricultural Benefits and Environmental Changes Resulting from the Use of Digested Sewage Sludge on Field Crops,* Environmental Protection Agency (Washington, D.C.: U.S. Government Printing Office, 1971), pp. 1–62.

[55] Carl W. Carlson and J. D. Menzies, "Utilization of Urban Waste in Crop Production," *BioScience,* June 15, 1971, Vol. 21, 561–64.

[56] As example of research, see Louis T. Kardos, "A New Prospect," *Environment,* March 1970, Vol. 12, No. 2, 10–21 and 27.

[57] *Cost of Clean Water: Cost Effectiveness and Clean Water,* p. 35 (see [4]).

Overfed

Eutrophication is an insidious form of water pollution that causes progressive deterioration of water resources on a wide scale. Overenrichment with the very nutrients that are essential for aquatic life results in an overabundance of plant life that is followed by other drastic changes in the ecology of lake, stream, or estuary. Although the full extent of eutrophication in the United States is unclear, a 1971 EPA survey provides an indication of the scope: Questionnaires sent to officials in all states with instructions to list types and sources of lake pollution produced information on 425 lakes; in 80 percent of these, the major problem was overgrowth of algae or other aquatic vegetation or some unspecified eutrophic condition. In most cases in which the causes had been determined, the nutrients were in wastes from municipal sewage treatment plants, septic tanks, or urban drainage [1].

Eutrophication—Illustrated

The nature of eutrophication can be illustrated by a series of events in Lake Washington on the northwest coast of the United States. The lake covers approximately 38 square miles and has 50 miles of shoreline, 20 miles of which are occupied by the city of Seattle. That city and others began dumping raw sewage into the lake in the late 1800s. Because of health hazards in the ensuing years, primary treatment of all sewage deposited in the lake was required by 1941 and, subsequently, secondary sewage treatment was required as well [2]. In 1955, 6.4 million gallons of treated sewage were pumped into Lake Washington each day from ten plants [3]. Except for an occasional in-

put of raw sewage due to storm-sewer overflows, the health problem had apparently been solved. However, by the mid-1950s other problems had developed and, despite secondary treatment operations, increases in the algal population forced limitations on recreational uses of the lake.

Scientists have been studying the condition of Lake Washington for many years. They have monitored the amounts as well as the kinds of algae which occur there, the lake's oxygen concentration, its chemical characteristics, and its clarity. Although in both 1933 and 1950 varieties of algal species were observed, by 1955, *Oscillatoria rubescens*—a type of blue-green algae frequently identified with polluted lakes—was the dominant species. Not only the type but the number of algae changed. The total number of algal cells per milliliter of water was 935 in 1950, rising to 1725 in 1955. Increased algae decreased the penetration of light: Measured from the surface of the water, visibility was 3.7 meters in 1950, 2.1 meters in 1955, and 1.0 meter in 1963. Oxygen depletion in 1955 was substantially greater than in previous years. These were all indications that Lake Washington was in biological trouble [4].

Alarmed at the deterioration of their lake, the people of the area raised $100 million for a counterattack. In 1958, a government corporation, the Municipality of Metropolitan Seattle (Metro), was formed, and by 1968 all sewage effluents had been successfully diverted [5]. Diversion was simply a euphemism for piping the sewage discharges away from the lake and dumping them into Puget Sound, a salt water inlet from the Pacific Ocean. After this technological sleight of hand, the Lake Washington pollution problem again seemed to be solved. The biological condition of the lake improved and

Matthews Beach on Lake Washington, full of happy swimmers in 1958.

swimming, boating, and fishing enthusiasts no longer had to worry about troublesome scum and odor problems. Scientists studying the lake documented the improvements. In 1969, the summer light penetration was 2.8 meters—not yet back to the clarity of 1950, but a vast improvement over 1963. Phosphates showed a greater decrease than nitrates: Phosphates in 1969 decreased to 28 percent of their 1963 concentration, nitrogen to 80 percent of its previous value. Because of the close correlation between the decline in algae and the decline in phosphates, biologists concluded that removal of phosphates otherwise present in effluent even after secondary treatment was in large part responsible for the improved condition of the lake [6].

Other Considerations

Lake Washington's "success" story, however, raises several fundamental questions. First of all, at the federal level most of the money and effort to be expended by the government on water pollution control in the next few years will be focused on secondary treatment facilities throughout the nation [7]. Will the standards employed in this type of waste treatment facility be adequate in light of the problems encountered in Lake Washington, which, despite secondary treatment for many years, produced serious eutrophication?

Five would-be swimmers in dry trunks on the same beach in 1968. Since that time, Matthews Beach has again become safe for swimming; the sewage that made the lake unhealthy and unsightly is now treated and diverted to Puget Sound.

As a related question, what specific nutrient-bearing pollutants caused the eutrophic deterioration and what can be done to prevent them from reentering the environment if secondary treatment cannot? Then, are the conditions alleviated in Lake Washington reversible in all lakes? That is, are the factors which may destroy ecological balance the same for all lakes, as well as for other bodies of water such as streams and estuaries? This consideration is important in determining the strategy and allocation of funds for pollution countermeasures.

Finally, one must consider that from 1963 to 1968, 20 million gallons per day of treated sewage effluent were routed away from Lake Washington into Puget Sound, adding to the burden of sewage already being dumped there. Since 1970, all the sewage going into the Sound from Seattle has received primary or secondary treatment (in 1961, 70 million gallons per day of raw sewage was discharged into the Sound). But even treated wastes, as we have seen, can create eutrophic conditions. Some experts believe that these conditions will not develop in Puget Sound since it has always been rich in naturally occurring nutrients, yet has never become eutrophic because of the frequent replenishment and mixing of the water by tidal action. Toxic pollutants in the treated sewage are another problem, however. Metro can refuse to accept effluents that contain undesirable amounts of toxic elements, and has done so in the case of zinc from the photoengraving laboratories of Seattle newspapers; faced with removing the zinc or changing their engraving process, the newspapers elected the latter [8].

Choosing to divert wastes to the Sound to alleviate the eutrophication in Lake Washington means that Seattle must now carefully monitor industrial effluents to catch toxic wastes before they build up to dangerous levels as well as continuously study the effect of the sewage on the receiving waters and their aquatic life. The diversion of sewage from one body of water to another can only delay solving the ultimate problem of finding ecologically sound ways of disposing of wastes and raise further questions about our overall waste treatment strategy.

Treatment Strategy

There are three types of sewage treatment in use today—primary, secondary, and tertiary. Primary and secondary treatments, however, are the major types and shall be discussed here.

Primary Treatment. Primary treatment, the most elementary, is a physical process in which sewage is piped to open tanks where the solid wastes are allowed to settle. Particles that float to the surface may be skimmed off, but the remainder of the sewage is discharged directly into the stream. Bacteria and fungi proliferate at sewage outfalls since they utilize sewage as well as plant and animal remains for food. These microorganisms are responsible for the decomposition of the waste matter, securing energy from the organic substances (compounds containing carbon) and releasing the inorganic substances, including nitrogen and phosphorus, into the stream again. While this activity, from an ecological point of view, is important in returning nutrients to the food cycle for utilization by higher plants and animals, bacteria must consume a great deal of oxygen in order to make use of the available organic compounds.

Just below the sewage outfall oxygen may be depleted completely. Here anaerobic bacteria (bacteria which can live without oxygen) carry out the decomposition process. These bacteria release foul-smelling methane and hydrogen sulfide, instead of the carbon dioxide released by oxygen bacteria. At greater distances from the point of effluent release, oxygen content increases as the wastes become more diluted.

The sludgeworm is frequently found in water with low oxygen content and is an indicator of organic pollution. These worms can tolerate oxygen concentrations as low as 0.5 ppm. Much higher concentrations of oxygen, however, are necessary for a balanced aquatic ecosystem. For a diversified warm-water fish population which includes such fish as bass and sunfish, the

The sludgeworm is one biological indicator of polluted waters with low oxygen content. Pockets of high sludgeworm populations in the Lake Michigan Basin are shown in this map, numbered per square meter.

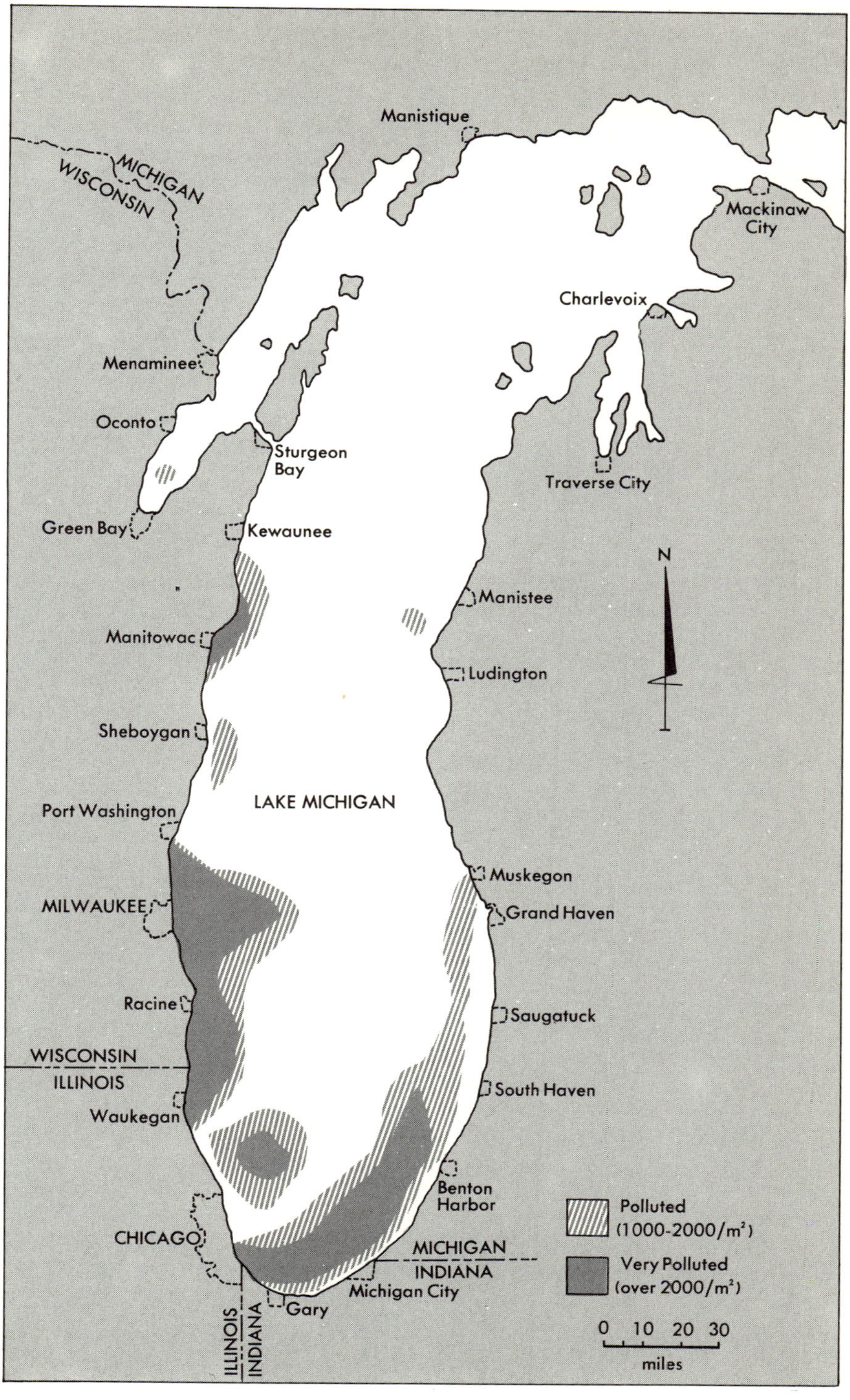

Manistique
MICHIGAN
WISCONSIN
Mackinaw City
Charlevoix
Menaminee
Oconto
Sturgeon Bay
Traverse City
Green Bay
Kewaunee
N
Manistee
Manitowac
Ludington
Sheboygan
LAKE MICHIGAN
Port Washington
Muskegon
MILWAUKEE
Grand Haven
Racine
Saugatuck
WISCONSIN
ILLINOIS
South Haven
Waukegan
Benton Harbor
Polluted (1000-2000/m²)
CHICAGO
MICHIGAN
INDIANA
Very Polluted (over 2000/m²)
Michigan City
Gary
ILLINOIS
INDIANA
0 10 20 30
miles

FWPCA criteria is a dissolved oxygen concentration above five ppm that may drop to four ppm for only short periods of time. Cold-water fish like salmon and trout require even more oxygen: The FWPCA criteria is seven ppm, and may not drop below five for more than six hours at a time; in the hypolimnion—the lower, colder layer of water where trout live—the dissolved oxygen concentration should never drop below six ppm [9].

Secondary Treatment and BOD. In order not to overburden nature's ability to consume organic matter (and to prevent oxygen depletion), secondary waste treatment may be provided. *Secondary treatment,* in effect, mimics natural processes. Sewage is incubated in large tanks where bacteria are allowed to consume approximately 75–90 percent of the organic matter before the effluent is released. The stream will presumably be able to incorporate the remaining 10–25 percent of the organic substances without becoming devoid of oxygen.

To determine what effect sewage effluent from secondary treatment plants will have on the oxygen concentration of the receiving stream, sanitary engineers determine the effluent's BOD—its biological (or biochemical) oxygen demand. The BOD is the oxygen required by bacteria to decompose organic wastes. Samples of treated sewage effluent are incubated in laboratory flasks for five days at 20°C. The difference between the amounts of dissolved oxygen present at the beginning and at the end of these incubation periods (or the amount of oxygen lost) is the BOD. From this test it is then possible to determine whether organic wastes will reduce the oxygen in a stream below the level critical to sustain aquatic life.

The convenience of the BOD measurement has made it the favorite tool for assessing waste treatment effectiveness and even general water quality. Summarizing the EPA Water Quality report:

> Five-day biochemical oxygen demand (BOD_5) is probably the most useful general indicator of the strength of organic wastes; . . . it is the most widely used means of expressing, in almost shorthand fashion, the general quality of water; and it is accepted by sanitary engineers, if not ecologists, as a surrogate for other parameters in broad descriptions of waste characteristics or of stream quality. . . . Because of its well established position as the prime measurement of waste strength, BOD reduction is the standard indicator of waste treatment plant efficiency. [10]

In contrast, an earlier caution about BOD was sounded in a review prepared for the FWPCA, predecessor of the Water Quality Office. The author, F. W. Kittrell, commented:

> The efficiencies of sewage treatment plants were, and still are, cited in terms of BOD reductions, rather than bacterial or solids reductions—or something else equally neglected. . . . The simple truth is that, nearly a half-century after Streeter and Phelps first published the method, there are still so many unknowns in the relationship (between BOD and dissolved oxygen) and so many factors that are difficult to measure that its application in actual stream situations frequently is fraught with frustration, and even failure. [11]

Waste Treatment—A Source of Eutrophication

The problem in the secondary treatment system which appeared to be associated with eutrophication in Lake Washington could not be diagnosed by BOD measurements. Although the bacteria consume organic matter before it reaches the receiving water, thus preventing the depletion of oxygen in the stream or lake, this very process releases the inorganic substances which act as plant nutrients and are discharged into the stream. There, nutrients such as phosphates and nitrates feed the algae until they become overabundant and die. Decaying organic masses result, reducing the dissolved oxygen content of the stream just as much as the organic loads discharged from the older primary treatment facilities which these secondary treatment plants replaced.

In 1968, 467.1 million pounds of phosphates entered surface waters from all types of municipal sewage treatment facilities serving 139.7 million people [12]. In the same year, approximately 1.16–1.35 billion pounds of nitrogen entered U.S. surface waters from human excrement in effluents from primary and secondary sewage treatment plants. Human wastes are the main source of nitrogen in domestic wastes [13]. The approximate amount of nutrients removed by each type of treatment appears below:

Waste Treatment Process	*Primary*	*Secondary*
Total nitrogen [14]	15–20%	30–50%
Total phosphorus [15]	10%	30%

The oversight in secondary treatment strategy that promotes eutrophication is the result of utilizing one segment of the aquatic cycle and ignoring what follows. Thus, in addition to being concerned with the balance of oxygen between oxygen-producing photosynthetic green plants and oxygen-consuming animals and bacteria, the closely related balance of nutrients and other necessary growth factors must also be considered. In the process of photosynthesis, plants use carbon dioxide, water, and sunlight in order to produce carbohydrates. Essential plant nutrients also include a number of inorganic elements and compounds. Nitrogen is needed for the production of amino acids—the building blocks of proteins. It is available to algae in the form of nitrates (NO_3), nitrites (NO_2), and ammonia (NH_3). A limited source is nitrogen gas (N_2) which may be fixed (changed chemically into nitrates) and incorporated into the cells of certain bacteria and blue-green algae. Phosphorus is needed in small quantities in all living things to produce ATP (adenosine triphosphate), an important energy molecule, and the nucleic acids DNA and RNA which transmit genetic information. Phosphorus is supplied to plants mainly in the form of phosphates which, under natural conditions, are derived from phosphate-bearing rocks such as apatite.

Nitrogen and phosphorus are both found in a variety of forms in aquatic ecosystems as the result of domestic and agricultural wastes. Untreated sewage contains nitrogen, usually in the form of urea and ammonia, which may be converted in some secondary sewage treatment processes, such as the activated sludge method, to nitrates as a result of biological activity. Phosphates are also found in sewage in various forms [16]. Some of the insoluble, complexed forms, such as those phosphates adsorbed onto suspended solid particles, are removed by primary and secondary treatment. The soluble phosphates, like polyphosphate (from detergents) and orthophosphate (from the breakdown of detergents and human wastes), are discharged into the aquatic environment and provide food for aquatic plants.

The Life of a Lake

Ecologists often classify lakes according to the amount of nutrients which they contain, generally dividing them into two major categories—oligotrophic and eutrophic. An *oligotrophic* lake has few nutrients and, as a result, supports small plant and animal populations. It contains a variety of species and a good supply of oxygen in its bottom waters. Eutrophication, on the other hand, refers to "the process of enrichment with nutrients" [17], thus a eutrophic lake contains large plant and animal populations.

Most lakes support natural cycles of algal growth or "blooms." There are usually two major blooms—one in the spring and one in the fall—due presumably to the injection of nutrients into the photosynthetic zone during spring and fall turnovers. More frequent algal blooms and a limited diversity of species are both indicators of eutrophication. Blooms often consist of certain types of blue-green algae and, because of the oxygen depleting effect of eutrophication in bottom waters, only those species that can tolerate low oxygen concentrations can survive. In Lake Erie, for example, both may flies and caddis flies (food sources for many fish) have disappeared, while the number of sludgeworms has increased tenfold [18]. Thus, although Lake Erie is often described in the popular literature as being "dead," it now contains a greater biomass than ever—although the species present are of a less desirable quality.

There is some debate in the scientific literature concerning the process by which a lake may become eutrophic, an argument that has some bearing on attempts to understand natural as opposed to man-made eutrophication. It has been widely thought that lakes are oligotrophic when formed, become eutrophic with time, and finally, due to the accumulation of sediments bearing dead algae, fill up, age, and eventually disppear. Some evidence, however, has shown that, under natural conditions, lakes are either oligotophic or eutrophic at the time of their formation, depending on the nutrient content of the watershed, and remain that way, evidenced by the fact that some lakes have remained oligotrophic since their formation thousands of years ago. Before the 1900s, Lake Erie was an oxygen-rich, clear lake and probably was not aging [19]. At any rate, it is beyond dispute that today's wastes drastically

Dead and dying algae along a bank of Lake Minnetonka, west of Minneapolis, Minnesota.

speed up the eutrophication process. If, as is usually the case, these wastes include large amounts of nutrients, an oligotrophic lake may become a eutrophic one and the degree of eutrophication in an already eutrophic lake may increase.

The effect of these wastes on the eutrophic conditions of lakes has been well documented. A 1947 article summarized the condition of over 309 lakes in Europe and North America which had become eutrophic [20]. An interesting example cited was that of the Zurichsee in Zurich, Switzerland. This lake is composed of two distinct basins, the Untersee (141 meters deep) and the Obersee (50 meters deep). The deeper basin received urban sewage effluent and was transformed in a short period of time (about 50 years) from a clear, oligotrophic lake to a eutrophic one, as evidenced by a change in fish populations, the number of algal blooms, and other factors. The shallower of the two lakes did not receive sewage during this time period and was not affected by eutrophication. This represents a sort of inadvertently controlled, scientific experiment which demonstrates human involvement in the eutrophication process.

The size of a body of water does not exempt it from the influence of added waste. Two well-known cases of the massive disruption of an aquatic environment have occurred in Lake Erie and in the Baltic Sea [21]. In both instances, natural conditions—the shallowness of Lake Erie and the Baltic's long periods of stagnation—make excessive inputs of nutrients extremely hazardous to the biological health of the environment.

To uncover the basic causes of eutrophication requires a consideration of the factors involved in disturbing the equilibrium of the ecosystem. First of all, certain lakes, because of their physical or chemical characteristics, are much more likely to be adversely affected by overfertilization. In deep lakes, for instance, a barrier develops between top and bottom waters because of thermal stratification: Surface water heated by the sun is less dense than the cold layers below, so the less dense layer floats on top of the denser layers in the same way that oil floats on water. This division creates a definite barrier against the transport of dissolved materials from one layer to the other. If no mixing ever occurred, the biological activity necessary to break down the dead plants and animals that sink to the cold bottom layer would completely deplete the oxygen there. Nutrients trapped in this hypolimnion, if never mixed with upper layers, would be unavailable for the growth of oxygen-producing plants at the surface which in turn help to renew the reduced dissolved oxygen supply. This potential barrier to biological activity is usually overcome naturally in the spring and fall each year. At those times, essentially complete vertical mixing occurs, redistributing oxygen and nutrients and permitting normal biological cycles to continue. However, in a shallow lake, such as the western basin of Lake Erie which is only 6.9 meters deep, there is less thermal stratification, nutrients are always available, and algal blooms may become much more frequent. As more and more algae die and sink to the bottom, the lake becomes more and more shallow. The rate of eutrophication will increase significantly when the lake becomes shallow

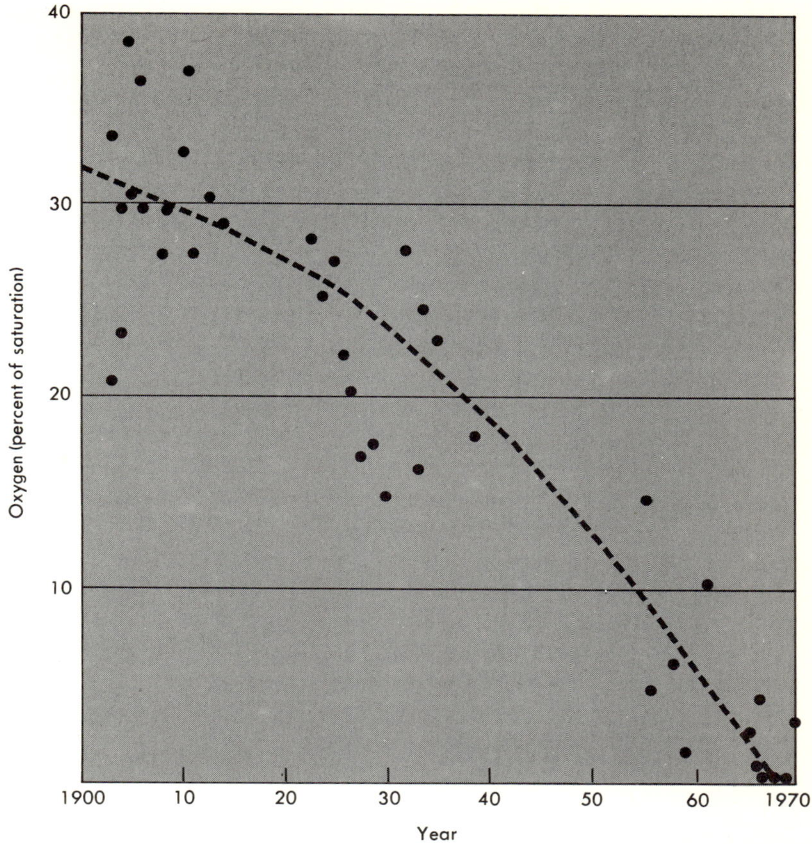

From 1900 to 1970, there has been a precipitous drop in dissolved oxygen in the North Central Basin of the Baltic Sea, which has been used as a dump by bordering nations. Natural barriers between the sea and the Atlantic Ocean tend to make the Baltic stagnant and therefore a highly inappropriate receiving water for waste.

enough for light to penetrate to the bottom, allowing large aquatic weeds to grow. As these plants die, sediments may accumulate very rapidly.

Limiting Factors

In studying the causes of eutrophication, it is necessary to learn not only how nutrients are distributed in the aquatic environment, but which of the many nutrients needed for plant growth are most important. Plants do not need identical quantities of all essential nutrients. The most critical nutrients are usually required in a ratio of 106 units of carbon to ten of nitrogen and one of phosphorus. If only one of these elements is missing, the others become useless to a plant; if one is found in very small quantities, even though the other nutrients are available in excess, the limited element will determine how much of the other nutrients can be used. If only half a unit of phosphorus is present in the water for every ten units of nitrogen and 106 of carbon, only half the nitrogen and carbon will be used by the algae or other aquatic plants. The presence of large amounts of carbon and nitrogen in the water and the absence of phosphorus would indicate that phosphorus was the factor limiting plant growth. If sewage effluents rich in phosphorus were added to this water, an increase in plant productivity could result.

To understand the limiting factor concept in relation to the three critical nutrients, consider the following hypothetical analogy. To purchase a package of cigarettes from a coin-operated dispenser, presume you *must* insert into three appropriately sized slots a nickel, a dime, and a quarter. If you are missing one coin, even the humble nickel, the other two are useless. You cannot operate the machine even though you may have three dimes and two quarters. Similarly, if phosphorus, for example, is missing in a body of water, an aquatic plant cannot utilize any of the nitrogen and carbon present, even if the latter elements exist in abundance. The two nutrients generally considered by scientists to be the most frequent limiting factors in the aquatic environment are the nitrates and the phosphates. Phosphates, in particular, are regarded by some scientists as the major limiting factor because their levels are so low in nature in relation to plant needs [22].

A number of studies have been conducted in an effort to determine the sources of nitrogen and phosphorus and the relative amounts of each which can be assigned to natural sources as compared with agricultural, industrial, or domestic sources. Lake Mendota in Madison, Wisconsin, was the site of one extensive survey. There, 36 percent of the phosphorus was from municipal and industrial waste; 17 percent from urban runoff, which included lawn fertilizers, automobile exhausts, precipitation, and chemicals used in street deicing; 42 percent from rural runoff from croplands, woodlands, pastures, manure, and drainage from silos and barns; two percent from precipitation; and two percent from ground (subsurface) water. Ten percent of the nitrogen entering the lake was from municipal and industrial waste; six percent from urban runoff; 11 percent from rural runoff; 20 percent from precipitation; 52 percent from ground water; 0.4 percent from nitrogen fixation; and some fraction of the nitrogen from several sources was attributable to contamination from nitrate fertilizers. Small but undetermined amounts of both phosphorus and nitrogen were also attributed to marsh drainage. From these estimates the authors of this study concluded that 50 percent of the nitrogen and 90 percent of the phosphorus in Lake Mendota was associated with wastes from human activities, added intentionally or unintentionally to the lake [23]. Only 12 percent of the nitrogen and 56 percent of the phosphorus entering Lake Washington in Seattle in 1957 was from sewage effluent [24]. Different studies reflect some variations in the proportion of nitrogen and phosphorus assigned to their respective sources. Variations would be expected according to the degree and type of agriculture and industry, the size of the city, and the type of waste treatment in the test area, although only

Agricultural runoff carries nutrients added to the soil to increase crops into rivers and lakes where they increase crops of algae instead.

further study will resolve some of the interpretive differences. The sources of the nutrients in the studies, however, were predominantly human activities.

Two of the more readily identifiable man-made sources of nutrients are the inorganic nitrogen fertilizers used extensively in agriculture and the phosphates from sewage effluents, a large part of which are from detergents. Much of the nitrates in surface water have been attributed by some scientists to the increasing use of inorganic nitrogen fertilizers. Studies carried out in the Sangamon River Watershed in central Illinois in the spring of 1970 showed that at least 55–60 percent of the nitrate found in surface waters there came from fertilizer nitrogen rather than from natural sources [25].

Limiting the amount of nitrate entering the water is important not only because it may contribute to eutrophication but because of a health hazard, methemoglobinemia, to be discussed later. For these reasons, regulation of fertilizer is now being considered. Robert Harmeson of the Illinois State Water Survey pointed out at state hearings on the subject that the nitrate concentration of the Illinois River and most of the watersheds studied has been steadily increasing from 1897 to 1971 [26]. Since phosphates are also used in fertilizers (in the United States, 80 percent of phosphate supplies are used in fertilizers [27]), this factor also was discussed at the hearings. It was reported that the amount of phosphates from fertilizers which may increase the phosphates in surface waters is difficult to determine because, unlike nitrogen, phosphorus is more readily adsorbed to soil particles and does not move into ground water. Much, however, may be contributed by soil erosion [28].

Controversial Issues

The ecological controversy surrounding the question of which nutrient is responsible for eutrophication has become a political and economic issue as well. Consideration has been given to the removal of phosphates either from sewage effluent or from detergents. The first suggestion, considered in more detail presently, has been made because technological methods for the removal of phosphorus, unlike those for nitrates, are available; the rationale for the second suggestion is that approximately 30 percent of all phosphates found in surface waters come from detergents. A number of state and local governing bodies have moved to ban the use of detergents containing phosphates. At the federal level, some congressmen prefer to see phosphates banned from all detergents—a suggestion that has naturally met with much opposition from the soap and detergent industry. A report from the Committee on Government Operations, based on April 1970 hearings before the Conservation and Natural Resources Subcommittee concerning phosphates and detergents, summarizes the scope of the arguments [29]. Industry's first claim is that phosphorus has not been proven to be the cause of eutrophication since phosphorus is not the only nutrient necessary to nor is it always the limiting factor in eutrophication. This argument is based primarily on a number of articles written by L. E. Kuentzel, a scientist associated with the Wyandotte Chemical Company (one of 20 major polluters of Lake Erie) [30], which state that bacteria grow symbiotically with algae, with the bacteria providing the carbon dioxide from which the algae obtain carbon. Thus, in some cases, carbon dioxide rather than phosphorus may be the limiting factor, according to Kuentzel's argument. A second line of reasoning taken by the soap and detergent industry is based on an article written by C. N. Sawyer, whose study of a number of Wisconsin lakes showed that an increase of phosphorus above .010 mg/l caused eutrophication in those lakes. From this, the industry concludes that adding *more* phosphates to an *already* eutrophic lake will not further harm the situation since there is already more than enough material available for algal growth. The related argument by industry is that *removing* phosphates from detergents won't help since there is so much eutrophication already.

Scientists who testified at the congressional hearings, as well as other scientists working in the field, contest the points raised by proponents of phosphate detergents. In response to the question of whether or not phosphate is the limiting factor, it was agreed that phosphate is not the only nutrient necessary for plant growth and that the limiting factor may change in any body of water from time to time as the nutrient input changes and in turn changes the ratio of nutrients in the water. However, studies have provided good evidence that phosphorus is the limiting factor in a number of lakes.

The conclusion that phosphorus is limiting in Lake Washington under natural conditions was reached on the basis of studies of the lake made in various years. In 1933, when the lake was relatively unpolluted, nitrates were left over once the phosphates had been used up by the plants, indicating that phosphates were the limiting factor and that their addition to the lake could stimulate algal growth. In 1962, however, the reverse was true: Phosphates remained once the nitrates had been consumed, reflecting the immense input of phosphates in sewage. By 1968 and 1969, after sewage diversion, nitrates were again in excess and phosphates were the limiting factor [31]. Phosphorus is also thought to be the limiting factor in Lake Erie. This opinion is based on laboratory experiments reported in the International Joint Commission report of Lake Erie [32] in which samples of lake water were supplemented by varying amounts of phosphates and nitrates and the growth of algae was monitored. Lake Minnetonka, a large lake in Minnesota, has been studied extensively and it was found that phosphorus is limiting there in the summer [33].

On the other hand, in some cases nitrates may be the limiting factor in fresh water: nitrogen, like phosphorus fertilizers, is sometimes added to lakes and fish ponds by fish growers in order to stimulate algal growth and increase fish populations [34].

Nitrogen may also be the limiting factor in coastal waters and estuaries, as demonstrated in Great South Bay and Moriches Bay on Long Island, New York. In both bays, large amounts of nitrates and phosphates were released in the effluent from duck farms, resulting in algal blooms. Two types of experiments were conducted to determine which nutrient was the limiting factor. Water samples from areas with dense algal growths were analyzed for phosphate, nitrate, nitrite, ammonia, and uric acid concentrations. Phosphorus was found, but, except at the source of the effluent, no nitrogen was present, suggesting that nitrogen was the limiting factor. Laboratory tests in which samples of water from the bays were enriched with either nitrogen or phosphorus were also conducted. Heavy growths of algae resulted from the addition of nitrogen, but not from the addition of phosphorus. The results of these experiments, and the fact that the blue-green algae found in estuaries do not readily fix atmospheric nitrogen, indicated nitrogen was the limiting factor in coastal waters [35].

Industry's contention that carbon dioxide is the limiting factor can be correct only if all other nutrients are already in excess. In one study, carbon dioxide was shown to be the limiting factor for algal growth in certain lakes in Georgia [36], but these results cannot be generalized for other lakes. It has been pointed out that in Lake Erie 100–120 mg/l of bicarbonate ions provide a ready supply for the natural formation of new carbon dioxide which is then available for plant growth [37]. Since phosphate levels have increased while carbon dioxide and bicarbonate levels have remained the same in the last 100 years, phosphates remain the most likely limiting factor in Lake Erie.

Critical Nutrient Levels. After determining the limiting factor, it is then necessary to pinpoint the critical level of that nutrient which will cause overfertilization. Although Sawyer found in his study of Wisconsin lakes that 0.010 mg/l of phosphate were necessary to sustain algal blooms, actually no "magic number" can be applied to all bodies of water. Algae have different nitrate and phosphate requirements, and the degree of eutrophication depends upon the species of algae in question as well as on the nutrient level. Some algae can store five to ten times the amount of phosphorus they need, so the phosphorus concentration in the water at any one time is a poor indicator of potential growth [38]. Experiments have revealed the fallacy in relying on industry's suggestion of a 0.010 mg/l critical level for eutrophication. In one set of experiments, various dilutions of effluent from treatment plants in the District of Columbia were added to the Potomac River. The level of phosphorus collected in the river was 0.037 mg/l. Blue-green algae growth resulted from the addition of sewage effluent in proportion to the amount of phosphorus in the effluent sample. These results showed that even in the face of already high levels of phosphate, the addition of more phosphate compounds further intensified algal growth [39].

Reversing the Process

Nutrient Reduction. A further question is whether or not man-made eutrophication can be reversed. Do eutrophic lakes now contain so much phosphate that removal of phosphates from detergents will have no remedial effect whatsoever? This question has unfortunately been confused by the conviction on the part of many that, since lakes age, once a particular state is reached in the lake's history, reversal is impossible. There is some evidence, however, that the condition of many lakes may be improved, although they may not be restored to an original and pristine state. Lake Washington is one example. Although diversion is a questionable means of stemming water pollution, results do illustrate that eutrophication can be reversed and that the health of such a lake can be restored. Diversion was also successful in three Wisconsin lakes—Lake Monona, Lake Waubesa, and Lake Kegonsa. These are naturally eutrophic lakes, but sewage effluents intensified their algal blooms. In order to curb this trend in Lake Monona, treatment plants were built in 1914 and effluents were transferred into Lake Waubesa. By the 1950s, effluents were bypassing all three lakes. The amount of copper sulfate (a herbicide) needed to kill algae in Lake Monona has dropped significantly—from 70,000–80,000 pounds in the 1920s and 1930s to only a few thousand pounds in 1955. Although farm drainage still causes some eutrophication, nuisance algae have decreased [40].

Would similar diversion tactics help Lake Erie? According to the International Joint Commission's report on Lake Erie and testimony given at the phosphate and detergent hearings, there is some chance of restoring the lake by phosphate removal. About 70 percent of the phosphorus reaching Lake Erie and Lake Ontario comes from municipal and industrial wastes, and detergents are the source of 50–70 percent of municipal and industrial phosphates. This means that detergents are responsible for close to half of all phosphate waste, and suggests that the removal of phosphates from detergents would decrease the phosphate level from 40 micrograms per liter (μg/l) of lake water to approximately 20 μg/l. Complete removal of phosphates from all sewage would reduce the phosphate concentration level in Lake Erie to 11.9 μg/l (lower than the 1942 level of 14 μg/l) [41].

The extent to which restrictions on the use of phosphate detergents relieve the influx of phosphate nutrients is illustrated by data supplied to the Subcommittee on the Environment of the Senate Commerce Committee in October 1971 by Councilman Carl Levin of Detroit. Councilman Levin reported on the change in phosphate levels in sewage entering Detroit's water tereatment facilities after phase I of Detroit's ban on the sale of detergents containing more than 8.7 percent phosphorus by weight went into effect July 1, 1971. These figures are listed below:

Date	*Pounds per Day*
1970	
July	58,500
August	55,200
September	54,400
1971	
July	42,000
August	47,000
September	42,600

It is expected that the changes will be more dramatic after phase II of the phosphate ban begins, when all phosphorus in detergents will be eliminated [42]. In Erie County, New York, the ceiling of 8.7 percent placed on the phosphate content of detergents in 1971 was reduced to 0.05 ppm in 1972. This resulted in more than a 60 percent reduction in orthophosphates in six treatment plants monitored, more than a 45 percent decrease in phosphates in the streams below these plants, and a 43 percent improvement in stream quality in terms of reduced BOD, turbidity, and algal blooms [43].

Unfortunately, a store of phosphates has already built up over a period of years in the bottom sediments of Lake Erie, so that sufficient nutrients to perpetuate an algal bloom can be released even if future sewage effluents are relatively free of phosphates. Nutrients stored in bottom mud are ordinarily prevented from mixing with the water by a protective layer of insoluble iron salts, but when the water is depleted of oxygen, this protective layer disperses. Studies at the Great Lakes laboratory of the State University of New York at Buffalo revealed that during those periods when the shallow western basin of Lake Erie stratifies, the bottom layer of the lake frequently becomes anaerobic (devoid of oxygen) as a result of the decay of algae that have died and settled to the bottom. Once the hypolimnion becomes anaerobic and the protective layer of iron salts breaks up, phosphates in the sediments are released 11 times faster than when oxygen is present. Many scientists are con-

Hospital wastes laden with phosphate-rich detergents trail into the East River in New York. Phosphorus has been shown to be the limiting factor in the eutrophication of some lakes.

cerned that with the anaerobic zone covering a wider area of the bottom and persisting for a longer period each year, the bottom sediments will become an important source of phosphates and thus of eutrophication [44].

At this point it would be difficult to predict how long the process of nutrient reduction in Lake Erie would take since there is such a great accumulation of phosphates in the lake sediments. Restoring the health of Lake Erie would also depend on the removal of many other pollutants currently discharged into the lake. The evidence of improvements in Lake Washington and the Wisconsin lakes is strong support for active measures to reduce nutrient input even in Lake Erie and other bodies of water where eutrophication is very advanced.

Additional Costs. A further reason for these reversal measures is that, in addition to the massive effects of eutrophication on the ecosystem and damage to the aesthetic and recreational value of water, eutrophication greatly increases the cost of sanitizing drinking water. Live or dead algae (especially the blue-greens) or fungi which grow on algae impart unpleasant tastes and odors to water and are sometimes toxic. Algae clog filters in treatment plants and, because of oxygen depletion, the iron, manganese, and sulfur held in complex under aerobic conditions are released and must be removed before the water is safe to drink. Dead algae may also color drinking water a pale yellow, necessitating the removal of the organic substances responsible. (Drinking water will be discussed in detail in the next chapter). When untreated water is used for agricultural purposes, toxic substances produced by blue-green algae may cause the death of cattle and sheep. The growth of weeds in irrigation canals poses a problem in very arid areas of the United States, since much water is lost from the canals by *evapotranspiration* (the loss of water which has passed through a plant and is evaporated at its surface) [45]. An unexpected cost of eutrophication is the damage to shoreline houses from hydrogen sulfide, the by-product of sulfur bacteria which grow in areas where oxygen is lacking. This foul-smelling gas reacts with lead-based paint to form a black precipitate which gradually discolors painted houses.

Preventive Measures?

How to prevent presently unpolluted lakes from becoming eutrophic and to reduce or reverse the situation in lakes already in advanced stages of eutrophication is still a matter of debate. As mentioned previously, proposed solutions range from attacks on the symptoms of the problem to alleviation of its root causes. One symptomatic solution is to dredge nutrient-rich sediments and dispose of them elsewhere. In some places, however, nutrients may be as deep as 10–15 feet below surface sediments, making dredging an imposing task. Then there is the problem of what to do with the dredging wastes. Often enough, the answer is to dump the wastes in the ocean, reminiscent of diverting sewage wastes into Puget Sound. At least two scientific studies have demonstrated that such disposal simply transfers eutrophication and toxicity problems from one body of water to another [46, 47].

Another method of symptomatic relief is the control of weeds with algicides and herbicides. This is not always a practical answer either. Recently, for example, the Tennessee Valley Authority (TVA) considered use of the herbicide 2,4-D to control watermilfoil, an aquatic weed in Melton Hill Lake—the only source of drinking water for the city of Oak Ridge, Tennessee, and one of the city's recreational centers. Because of the potential danger of this herbicide, found to cause birth defects in mice and often to persist in the aquatic environment for over six months, the proposed control program was abandoned [48].

The other major preventive proposal is to use various techniques to remove phosphates from effluents at municipal treatment plants. Of the techniques available, chemical treatment to precipitate or otherwise reduce the phosphorus content in waste water is usually regarded as the most practical, primarily because it is the least expensive. Two serious problems must be considered, however. First, phosphate-removing technology is applied in conjunction with either secondary or tertiary waste treatment, most effectively with the latter. This means that a municipality must first make the large investment necessary for at least secondary waste treatment, which has already absorbed large amounts of local and federal funds and which will require even larger future investments to bring to completion. Tertiary treatment—

A Pennsylvania State University experiment in irrigating forests and croplands with treated sewage is represented here. The first treatment removes suspended solids and decomposes the remaining waste. Instead of being dumped into a nearby stream (which formerly received the waste), the undrinkable waste water is now pumped by pipeline to areas of growing vegetation. At this stage, the water contains about six mg/l of nitrate nitrogen and about seven mg/l of phosphorus.

The treated sewage waste water is sprinkled on forested and agricultural areas, where it seeps down through the ground and is diluted by ground water. In the forest, nutrients from the waste water are cycled and recycled through the trees, the leaves, other vegetation, and the forest floor. In agricultural areas, nutrients from the sprinkled waste water are drawn into plants and carried off when crops such as hay are harvested.

When the water is drawn up through a well it is clean enough to drink and contains less than three mg/l of nitrate nitrogen and less than 0.04 mg/l of phosphorus. After use, the water travels through sewage pipes to the treatment plant. And the process continues . . . .

composed of more complex physical and chemical processes that remove toxic elements, nitrates, phosphates, or all three—is even more expensive and is used in less than one percent of U.S. waste treatment facilities. Furthermore, although chemical treatment for phosphorus removal is less expensive than other methods, it is still very costly. For example, the projected cost for phosphate removal by coagulation with lime and iron salts, including operating costs, is $43.10 per million gallons in waste treatment plants with a flow of from ten million gallons per day and $41.10 in plants with a flow of 100 million gallons per day (treatment costs are less at higher flow rates) [49]. Secondary treatment of domestic sewage without phosphorus removal was estimated in 1967 at $50 to $100 per million gallons [50]. To introduce this process, then, would force some treatment plants to spend nearly twice as much each year in operating expenses. Such a cost increase would require major

new municipal funding. And, finally, the problem of disposing of the extracted nutrients would still remain.

The second major problem is the extent to which chemical or other advanced treatment can prevent phosphates from entering the water. As we have already seen, municipal wastes contribute a large portion of the total phosphates entering surface waters. However, large amounts also enter the water from other sources such as industrial activities, agriculture, urban runoff, and private sewage treatment systems. Removal of phosphates in major metropolitan treatment plants therefore would take care of only part of the total problem. Furthermore, the amount of phosphates removed by processes varies. The preceding cost estimates for the use of lime and iron salts assume removal of at least 90 percent of the total phosphorus in municipal wastewater. Such a high removal rate has been achieved in pilot projects and in some small advanced plants now in operation, but the rate has been lower elsewhere. An average of 72–84 percent total phosphorus was removed by chemical treatment at two plants in Grayling, Michigan, and Mentor, Ohio, respectively [51]. The current goal in the Lake Michigan Basin is 80 percent removal.

Under some circumstances, removal of 80 percent of the total phosphorus load in the waste water may still leave enough of the nutrients in the waste effluent to perpetuate local eutrophic conditions, particularly in lakes already overburdened with inorganic nutrients. In sum, the strategy to remove phosphates in the waste treatment plant, rather than to eliminate or limit their use at the source in products such as detergents, fertilizers, and industrial waste liquids, leaves many loopholes which permit phosphate discharge to continue. In addition, the large expenditures required for this strategy places a heavy burden on the public taxpayer, who lately has had to respond to a variety of local and federal funding ventures in the field of waste control. Against this strategy, then, must be weighed the benefits of keeping phosphates and other contaminants out of the environment in the first place by eliminating their sources. Where sewage is returned to the soil, as in the "sewage farming" of some European cities, there is a double benefit: the need for phosphate fertilizer is reduced and the nutrients in the sewage contribute to the growth of useful crops instead of to the growth of algae [52].

Finally, the overreliance on useful but limited concepts such as BOD have helped to foster a waste control strategy that has serious flaws. In the case of eutrophication, the strategy overlooks critical facts in aquatic biology that frustrate the intent to maintain high levels of dissolved oxygen in the water. In the case of toxic wastes, to be discussed in the next chapter, the strategy concentrates immense resources in efforts to screen out dangerous materials released by human activities instead of preventing such release in the first place. Historically, the strategy has served to protect man from waterborne diseases, but the ways of man have changed, and the strategy is now being badly battered in attempts to meet new health threats produced by man himself rather than by nature.

[1] Martha J. Kefelle and Paul D. Uttormark, *Problem Lakes in the United States*, prepared for the Office of Research and Monitoring, Environmental Protection Agency, December 1971.

[2] Charles G. Wilber, *The Biological Aspects of Water Pollution* (Springfield, Ill.: Charles C Thomas, 1969), p. 143.

[3] W. T. Edmondson, "Phosphorus, Nitrogen, and Algae in Lake Washington after Diversion of Sewage," *Science*, August 14, 1970, Vol. 169, 690–91.

[4] W. T. Edmondson, G. C. Anderson, and Donald R. Peterson, "Artificial Eutrophication of Lake Washington," *Limnology and Oceanography*, 1956, Vol. 1, 47–53.

[5] "Metro—The First Ten Years. 1958–1968," a publication of the Metropolitan Council of Seattle, pp. 1–39.

[6] Edmondson, "Phosphorus, Nitrogen, and Algae in Lake Washington," p. 691 (see [3]).

[7] *The Cost of Clean Water, Volume II: Cost Effectiveness and Clean Water*, Environmental Protection Agency (Washington, D.C.: U.S. Government Printing Office, March 1971).

[8] W. T. Edmondson (Professor of Zoology, University of Washington, Seattle), personal communications, March 30, 1971 and August 28, 1972.

[9] *Water Quality Criteria,* report of the National Technical Advisory Committee to the Secretary of the Interior, Federal Water Pollution Control Administration, Washington, D.C., April, 1968, pp. 43–44.

[10] *Cost of Clean Water: Cost Effectiveness and Clean Water,* p. 25 (see [7]).

[11] F. W. Kitrell, *A Practical Guide to Water Quality Studies of Streams,* U.S. Department of the Interior, Federal Water Pollution Control Administration, 1969, p. 86. In 1925, the U.S. Public Health Service published the work of Streeter and Phelps who had made extensive studies of Ohio River water. They developed mathematical equations which made it possible to calculate the amount of oxygen depletion (or BOD reductions) at various points downstream from a pollution source.

[12] *Cost of Clean Water: Cost Effectiveness and Clean Water,* p. 34 (see [7]).

[13] Nava Narkis, "The Nitrogen Balance Through Sewage Treatment Plants," Soil and Aquatic Task Force, Center for the Biology of Natural Systems, Washington University, St. Louis, Missouri, unpublished paper, 1970, 7 pages.

[14] Ibid., pp. 1–7.

[15] *Cleaning Our Environment: The Chemical Basis for Action,* a report by the Subcommittee on Environmental Improvement, Committee on Chemistry and Public Affairs, American Chemical Society, Washington, D.C., 1969, p., 108.

[16] Richard D. Grundy, "Strategies for Control of Man-Made Eutrophication," *Environmental Science and Technology,* December 1971, Vol. 5, No. 12, 1186.

[17] Kenton M. Stewart and Gerard A. Rohlich, *Eutrophication—A Review,* a report to the State Water Quality Control Board, Sacramento, California, 1967, pp. 5–6.

[18] *Pollution of Lake Erie, Lake Ontario, and the International Section of the St. Lawrence River,* report to the International Joint Commission by the International Lake Erie Water Pollution Board and the International Lake Ontario-St. Lawrence River Water Pollution Board, 1969, p. 29.

[19] Barry Commoner, "Lake Erie, Aging or Ill?" *Scientist and Citizen* (now *Environment*), December 1968, Vol. 10, No. 10, 254–65.

[20] Arthur D. Hasler, "Eutrophication of Lakes by Domestic Drainage," *Ecology,* 1947, Vol. 28, 383–91.

[21] Stig H. Fonselius, "Stagnant Sea," *Environment,* July–August 1970, Vol. 12, No. 6, 2–11 and 41–48.

[22] G. E. Hutchinson, *A Treatise on Limnology,* Volume I (New York: John Wiley and Sons, 1967), in Chapter 12, "The Phosphorus Cycle in Lakes," p. 727.

[23] *Report on the Nutrient Sources of Lake Mendota* by the Nutrient Sources Subcommittee of the Technical Committee of the Lake Mendota Problems Committee, January 3, 1966 (revised April 15, 1969), pp. 28–33.

[24] Edmondson, "Phosphorus, Nitrogen, and Algae in Lake Washington," pp. 690–91 (see [3]).

[25] D. H. Kohl, G. B. Shearer, and Barry Commoner, "Fertilizer Nitrogen: Contributions to Nitrate Surface Water in a Corn Belt Watershed," *Science*, December 24, 1971, Volume 174, 1331–334.

[26] *Summary, Pollution Control Board Hearings on Plant Nutrient Proposals, Hearing No. 6*, held by State of Illinois Pollution Control Board, November 3, 1971, Urbana, Illinois.

[27] *Man in the Living Environment*, The Institute of Ecology Report of the Workshop on Global Ecological Problems, 1971, p. 57.

[28] *Chemical Week*, "Fertilizers Win First Round in Runoff Fracas," February 9, 1972, Vol. 110, No. 6, 11–12.

[29] *Phosphates in Detergents and the Eutrophication of America's Waters*, 23RD report by the Committee on Government Operations (Washington, D.C.: U.S. Government Printing Office, April 14, 1970).

[30] *Lake Erie Report: A Plan for Water Pollution Control*, U.S. Department of Interior, Federal Water Pollution Control Administration, Great Lakes Region, August 1968, p. 56.

[31] Edmondson, "Phosphorus, Nitrogen, and Algae in Lake Washington," pp. 690–91 (see [3]).

[32] *Pollution of Lake Erie, Lake Ontario, and the International Section of the St. Lawrence River*, p. 70 (see [18]).

[33] Herbert E. Wright and Robert O. Megard, "Lake Minnetonka: Nutrients, Nutrient Abatement, and the Photosynthetic System of the Phytoplankton," final report for the Office of Water Resources Research Project, University of Minnesota, December 1970, pp. 32–53.

[34] J. R. Vallentyne, "Phosphorus and the Control of Eutrophication," *Canadian Research and Development*, May/June 1970, Vol. 3, No. 3, 36–49.

[35] John H. Ryther and William M. Dunstan, "Nitrogen, Phosphorus, and Eutrophication in the Coastal Marine Environment," *Science*, March 12, 1971, Vol. 171, 1008–1013.

[36] G. E. Likens, A. F. Bartsch, G. H. Lauff, and J. E. Hobbie, "Nutrients and Eutrophication," *Science*, May 21, 1971, Vol. 172, 873–74.

[37] Vallentyne, "Phosphorus and the Control of Eutrophication," p. 37 (see [34]).

[38] Fred G. Lee, "Eutrophication," the University of Wisconsin Water Resources Center, Eutrophication Information Program, September 1970, pp. 9–10.

[39] *Phosphates in Detergents*, pp. 19–20 (see [29]).

[40] Arthur D. Hasler, "Cultural Eutrophication Is Reversible," *BioScience*, May 1969, Vol. 19, No. 5, 425–31.

[41] *Phosphates in Detergents,* p. 26 (see [29]).

[42] Councilman Carl Levin of the Common Council of the City of Detroit, statement to the Subcommittee on the Environment of the Senate Commerce Committee, October 29, 1971.

[43] Robert A. Sweeney, Director of the Great Lakes Laboratory, State University College at Buffalo, New York, personal communication, January 3, 1973.

[44] Ibid.

[45] Lee, "Eutrophication," pp. 23–27 (see [38]).

[46] M. Grant Gross, *Analyses of Dredged Wastes, Fly Ash, and Waste Chemicals,* New York Metropolitan Region Technical Report No. 7 of the Marine Sciences Research Center, State University of New York, Stony Brook, October 1970.

[47] *The Effects of Waste Disposal in the New York Bight—Interim Report for January 1, 1970,* The Sandy Hook Marine Laboratory, U.S. Bureau of Sport Fisheries and Wildlife, Sandy Hook, New Jersey, December 3, 1969.

[48] Eric Hirst and Harvey Bank, "Striking the Balance," *Environment,* November 1971, Vol. 13, No. 9, 34–41.

[49] Leon W. Weinberger, Federal Water Pollution Control Administration (now Water Quality Office), statement in *Conference Proceedings: Pollution of Lake Michigan and its Tributary Basin, First Session,* U.S. Department of the Interior, Federal Water Pollution Control Administration, February 1, 1968, Vol. 2:812–77.

[50] *Cleaning Our Environment,* p. 108 (see [15]).

[51] Clifford Risley, Jr., Federal Water Pollution Control Administration (now Water Quality Office), statement in *Conference Proceedings: Pollution of Lake Michigan and its Tributary Basin, Third Session,* U.S. Department of the Interior, Federal Water Pollution Control Administration, March 31, 1970, Vol. 1:109–110.

[52] Jonathan Allen, "Sewage Farming," *Environment,* April 1973, Vol. 15, No. 3, 36.

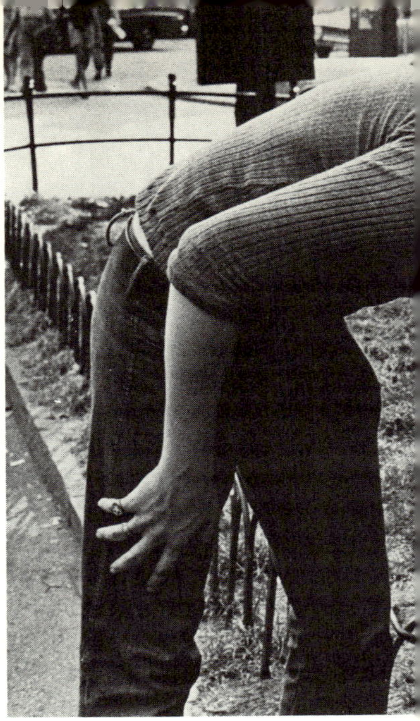

Drinking Water

In the 1880s, the death rate from typhoid in the United States was approximately 75–100 per 100,000 population. Today it is 0.1 per 100,000. Epidemics of typhoid fever and other waterborne diseases are no longer the principal health problems associated with water pollution. In their place are different disease microorganisms and a complex of human health hazards that can be related to the organic and inorganic substances dumped in the water in mammoth quantities by our fast-changing technological civilization. Some of these substances, among them a variety of chemicals and metals, pose a direct threat to human health of uncertain dimensions. Others, such as inorganic nutrients, threaten to degrade natural water quality, diminishing the availability of already precious drinking water and, in certain cases, promoting the growth of toxic algae and protozoans.

Before considering some of these new problems, a brief look at the old ones is in order, for we can by no means let down our guard against disease bacteria. In 1969, when the Bureau of Water Hygiene, a division of the Public Health Service (PHS), surveyed community water supplies serving more than 18 million people [1], the water in 120 of the 969 systems studied was found to exceed the limit for *coliform density* (the measure of bacteriological quality [2]. Most of the drinking water with excessive bacteria was found in small communities. This same survey found that 85 percent of the water systems investigated did not collect an adequate number of samples to maintain the regular surveillance necessary to detect potentially dangerous increases in coliform count. Forty-one percent of the systems failed to meet the PHS 1962 drinking water standards in one way or another: 25 percent exceeded certain

recommended limits and 16 percent exceeded certain mandatory limits. More than half the water treatment facilities examined had one or more major operating deficiencies. These included inadequate equipment or personnel to deal with disinfection, clarification (the removal of suspended particles), and the maintenance of proper pressure to ensure delivery of water.

Chemical Contamination

Appendix Table 2 summarizes the results of the PHS survey study of water in 969 communities. Sampling for chemical constituents in the study revealed that both the recommended and the mandatory limits for a number of chemicals were exceeded. Samples were collected from two or more points in each water supply distribution system and the results were then averaged. The PHS limits which were exceeded appear [3].

But the numbers in the table are average figures. What each family really wants to know is: What is coming out of *our* tap? Of course, it was impossible to individually sample the water supplies of 18 million people, but 2595 samples were taken directly from consumer taps and 36 percent of these contained at least one contaminant in an amount exceeding drinking water standards. By applying a factor for population equivalency, it was estimated that 2.9 million of the 18 million people surveyed were receiving substandard water. The heavy metals arsenic, iron, maganese, cadmium, and lead were found in some individual samples at concentrations exceeding the recommended or mandatory limit by ten times or more. Bacterial contamination as measured by the coliform count was also ten times the recommended standard or more in some samples, and so were suspended particulates as measured by turbidity units.

Ten percent of the individual samples exceeded the standard for total organic chemical compounds, but there were complaints by consumers of unpleasant taste and odor in water which was rated acceptable by the standard. Total organics are measured by filtering a sample of water through activated carbon and then using chloroform to extract the organics which have been absorbed on the carbon. However, not all the organics are absorbed onto the carbon and, of those absorbed, not all are extracted in the chloroform. There may therefore have been organics present which were not measured. PHS investigators also suggested that the unpleasant taste and odor might be caused by the interaction of the chlorine used in water purification with organic chemicals, even when the latter are present only in "acceptable" amounts.

Contaminants in open surface water (rather than in community water systems) were investigated at several hundred stations throughout the country by the U.S. Geological Survey in October 1970 [4]. Some stations were downstream from industries, some were on streams in undeveloped areas, and some were at the sources of drinking water supplies. The investigators tested for arsenic, cadmium, chromium, cobalt, lead, mercury, and zinc, and found all these heavy metals widely distributed in U.S. waterways in low concentrations and sometimes in amounts exceeding PHS drinking water standards. In two percent of the samples (from both undeveloped and urban areas), arsenic exceeded the limit of 0.05 mg/l; four percent of the samples, mainly near urban areas, were in excess of the limit for cadmium; and 63 percent of the samples contained 0.001–0.05 mg/l of lead, and a few exceeded the 0.05 limit. The recommended limit for zinc—5 mg/l—was also exceeded in a few cases. The relative importance of natural and industrial sources was not clear.

A more revealing picture of industrial pollution developed when the EPA investigated the lower Mississippi River [5]. In a 256-mile stretch of the river from north of Baton Rouge to south of New Orleans, a large concentration of modern chemical and petroleum refining works, including the largest petroleum refinery in the United States, has developed. The Mississippi serves as both a disposal site for industrial wastes and the source of raw domestic water for 40 utilities servicing about 1.5 million people.

Masses of dead fish, apparently killed by chemicals, tainted fish flesh, and drinking water with chemical odors and tastes prompted state officials to request the EPA's investigation of this area, which took place in 1969–1970. Contamination of drinking water was found and was traced to the discharge of huge amounts of pollutants from 60 major industries on the lower Mississippi. On a single day, each of 42 plants discharged at least one heavy metal to the river in amounts ranging upward from five pounds. The day's total discharges included 3700 pounds of lead from one plant, 71 pounds of cadmium from another, 396 pounds of copper from another, and 200 pounds of chromium from still another [6].

Each of these metals has well-known toxic properties. In the case of lead—

the third most prevalent waste in the industrial discharges—standards for short-term exposures have been set by the Public Health Service in addition to the mandatory limit of 0.05 mg/l considered physiologically safe over a lifetime. For a few weeks, as much as 2–4 mg/l of lead is not considered unsafe, although this same concentration borders on harmful if it continues for three months. Lead is toxic if taken in for several weeks at concentrations of 8–10 mg/l, and a concentration of 15 mg/l over the same period may be lethal—although the lethal concentration is not known with certainty.

How do 3700 pounds of lead dumped into the river per day from a single plant relate to these standards? This is difficult to say, since the EPA study did not test drinking water for lead. However, at the point where the effluent of one chemical company was dumped into the river, samples collected on three different days were found to contain 1.2 mg/l of lead on a day when 218 pounds of lead were dumped, 2.8 mg/l on a day when 485 pounds were dumped, and 4.8 mg/l on a day when 890 pounds were dumped. Dumping 3700 pounds of lead in a single day could therefore produce a concentration of lead in the effluent of as much as 20 mg/l. How much of this lead is still in the water when it flows out of kitchen taps in Baton Rouge, New Orleans, and the small towns along the Mississippi is one important question, but some of the lead could also reach humans via food chains.

Lead ingested from air, food, or water can accumulate in the body and cause serious damage to the kidney, liver, or brain, and to the reproductive and central nervous systems. At subtoxic levels, lead interferes with the functioning of red blood cells; at toxic levels, red blood cells may be destroyed. Children in particular are affected by lead poisoning—mental retardation, for example, is a common result of lead poisoning in children [7].

PHS standards for cadmium limit the concentration to 0.01 mg/l. A variety of health effects have been attributed to cadmium ingestion. Hypertension is correlated with increased retention of cadmium in the kidneys, and cadmium ingestion may cause cramps, nausea, vomiting, and diarrhea. The metal is perhaps most infamous as the cause of itai-itai disease, the hyperexcretion of calcium which leads to softening of the bones [8]. Cadmium was found in quantities greater than five pounds per day in eight different waste discharges in the lower Mississippi, but concentrations of cadmium in water samples were not reported.

Because of their known toxicity, the Public Health Service has set standards for other heavy metals found in industrial waste effluent in the Mississippi. These include chromium, copper, zinc, aresnic, and mercury, but again, as with cadmium, the data in the report are insufficient to relate the amounts found to the standards. (However, recall that the PHS study found concentrations of these heavy metals in excess of the standards.)

The same problem exists in interpreting the report's finding that 17 plants discharged ten pounds or more of phenols per day. Phenols impart an objectionable odor and taste to water, and may have been responsible, at least in part, for the "chemical" odor and taste in drinking water that triggered the Lower Mississippi study. The Public Health Service set a limit of 0.001 mg/l

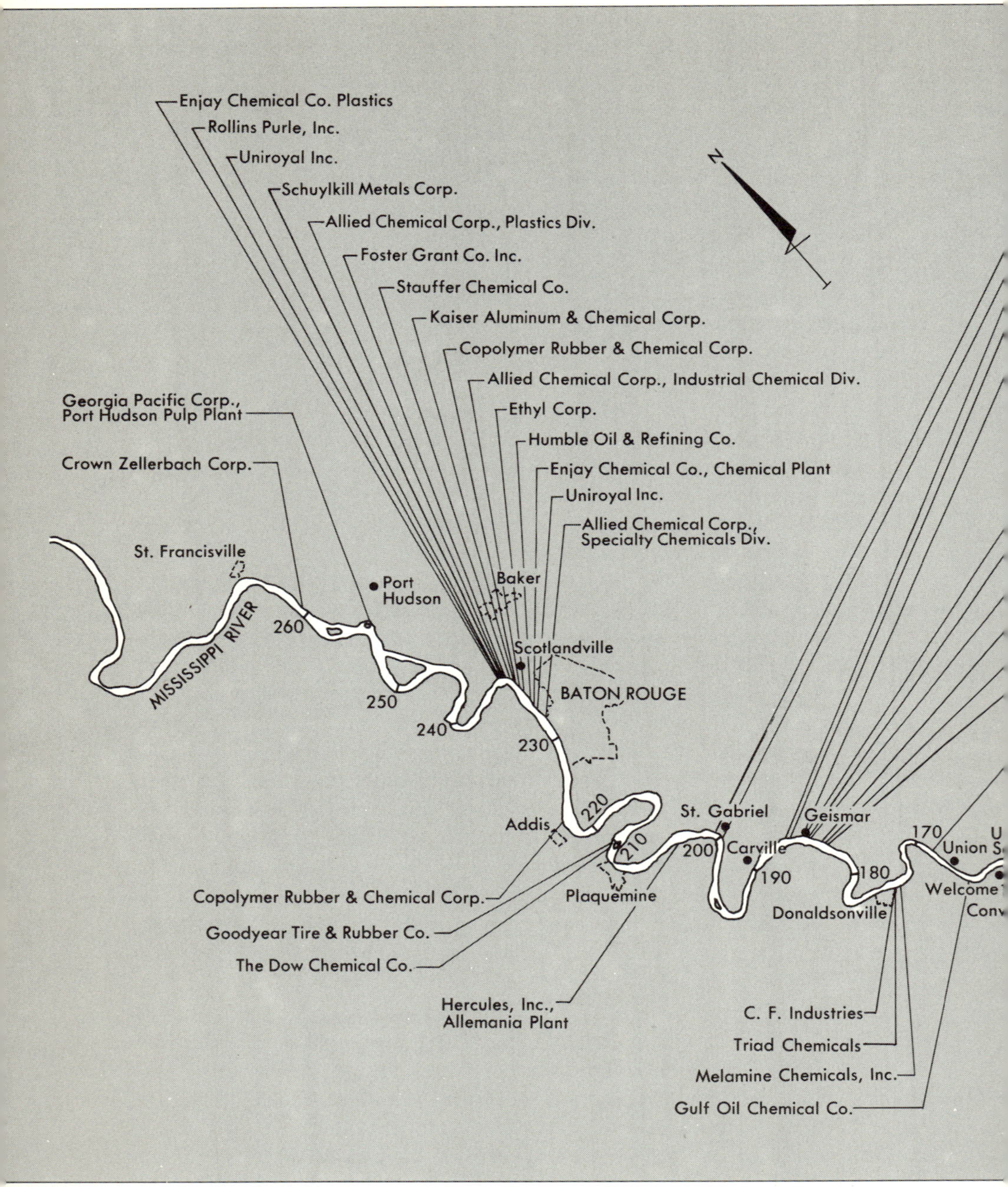

Hundreds of industries have found the Mississippi River a good location because it supplies them with water for manufacturing processes and a convenient dump for their wastes. This schematic representation of industrial polluters on the lower Mississippi River near new Orleans was prepared by the Environmental Protection Agency.

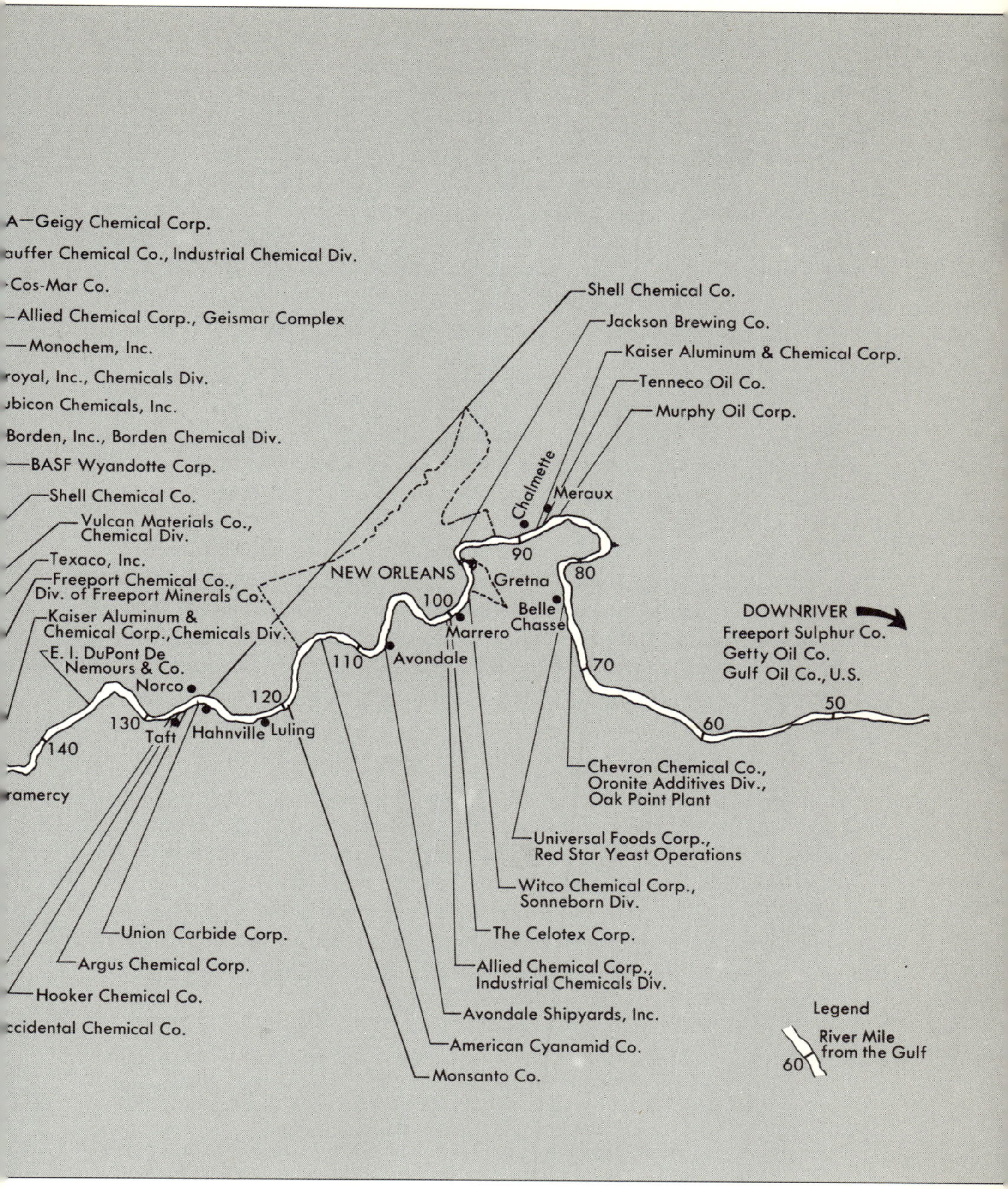

Of the 36 chemical compounds remaining in downriver drinking water after treatment, benzene, carbon tetrachloride, chloroform, dichloroethyl ether, ethyl benzene, and dimethyl sulfoxide have been found to be harmful to laboratory animals. How any of the 36 compounds, singly or in combination, affect people is not yet well known.

as the highest concentration of phenols permissible in drinking water, but because concentrations of phenols in water samples were not reported, it is impossible to say how the phenol discharge in pounds at the various plants relates to what appears in river water downstream or to what runs out of the taps in homes supplied with drinking water from the river.

Phenols are used extensively in the synthesis of organic chemicals, which accounts for their appearance in the effluent of chemical plants. The PHS permissible phenol concentration is set at a level intended to prevent taste and odor problems, and toxic effects are thought to occur only at far higher concentrations, partly because phenols are probably largely detoxified in the mammalian body. As much as 1000 mg/l—a million times higher than the standard—has been fed to rats with no observable effect [9].

Organic Chemicals. Many organic chemicals were also identified in the Mississippi discharges—88 in all were found either in streams receiving industrial effluent or in raw or treated drinking water supplies, or both. The concentrations of these chemicals were not measured. These *trace* organics were *estimated* to be present in parts per million or even in concentrations a thousand times lower—in parts per billion. Insignificant as these concentrations may appear, however, we cannot assume that even in such small amounts, singly or in combination, these chemicals are safe either for aquatic life or for human consumption. EPA analysts listed 36 organic chemicals found in water that had been treated at the Carrollton Water Plant in New Orleans and were ready for distribution to domestic users [10]. Some of these compounds, and others as well, were also found in tap water at the PHS Hospital at Carville, south of Baton Rouge.

In summarizing the results of the Lower Mississippi study, EPA scientists point out that while the Food and Drug Administration (FDA) has the authority to ban food additives shown to cause cancer, data on the carcinogenicity of contaminants in water has not been used in setting water quality standards. They also noted that toxicity tests are made on *healthy* laboratory *animals*. What the data derived means to people—some of whom are elderly, sick, or very young—has yet to be ascertained. The EPA analysts concluded:

> . . . there may be a myriad of organic chemicals, not yet isolated and identified, such as the pesticides, that could be present in these water supplies, which are carcinogenic, teratogenic, or mutagenic. . . . It is mandatory to single out potentially harmful substances and prevent them from entering those bodies of waters that will be used for drinking or food sources until it is clearly demonstrated that they pose no threat to human health or aquatic life [11].

Another EPA study related to these concerns involved a search of the literature to determine what organic chemicals are found in the water, what known industrial chemicals are probably in the water but have not been isolated yet, and what data are available on the possible health effects of both: toxic (poisonous to living things), carcinogenic (cancer-causing), teratogenic (causing abnormal offspring in the next generation), and mutagenic (harmful to future generations) [12]. It was determined that 66 organic chemicals reported in the literature have been found in fresh water and that 15 of the 33 of these tested for carcinogenicity caused cancer. Another 87 organic chemicals *suspected* of being water contaminants were tested for carcinogenicity, and 17 of these were found to cause cancer. Many of the same compounds were also mutagenic or teratogenic. In addition, 343 chemical compounds suspected of being water contaminants have not yet been investigated either to determine whether they are actually in drinking water supplies or to learn what their potential effects on health might be. Thus, 496 organic chemicals have been found to be or are suspected to be contaminating water in various parts of the United States. With new chemicals being produced every year, this number can be expected to increase—in fact, it has probably already increased since the EPA report was published in 1970.

In light of these hazards, how safe is water for the average consumer? The nationwide studies of drinking water that have been conducted have dealt primarily with contaminants regulated by the PHS 1962 drinking water standards, which generally have been incorporated into state standards as well. There are as yet no individual standards for any of the 496 industrial chemicals mentioned above, and the limitations of the standard for organics has already been noted.

The direct health threats of a number of chemicals, both organic and inorganic, can be expected to increase if present industrial and agricultural trends continue. Appendix Table 2 shows that 19 community water systems in the 1969 PHS survey contained excessive amounts of nitrate. These communities were not identified, but other studies indicate that the problem has arisen primarily in agricultural communities relying on wells or on river water that has traversed heavily fertilized areas as a source of water. The rise in the nitrate content of water in the Illinois farm country has already been mentioned. Increases have also been found in ground water in Nassau County, Long Island [13], and in the San Joaquin Valley, California [14].

Since World War II, the production of nitrogen fertilizers has increased fourteenfold and there has been a 70 percent increase in sewage nitrogen [15]. Fertilizers are probably the principal cause of the nitrates in drinking water, compounded in the San Joaquin Valley by the use of ground water for irrigation [16].

Nitrate. Nitrate as such presents no particular health problem, but nitrates may be converted by certain bacteria within the body to nitrites. The nitrites react with the hemoglobin in red blood cells, causing the cells to lose their ability to carry oxygen and thus threatening the body with asphyxiation. Infants are particularly susceptible to this condition—called *methemoglobinemia*—because their gastric juices have a low pH, which creates a suitable environment for the bacteria that convert the nitrates to nitrites, and because an infant's hemoglobin is more reactive with nitrites than an adult's. Mothers who are themselves unaffected by nitrate may pass it on to their infants in breast milk. Children may also receive nitrate from cow's milk if the cows drink water high in nitrate, from vegetables that have taken in nitrates from the soil, or from processed foods to which nitrate or nitrite has been added (if the food is not fresh or properly refrigerated, the nitrate may be converted to nitrite).

Since the link between nitrate intake and methemoglobinemia was first identified in 1945, more than 1000 cases of infant methemoglobinemia have been reported throughout the world, with 83 deaths. The more than 278 cases reported in the United States in the late 1940s were associated with high levels of nitrate in the water used in the infants' formula; 39 of these babies died [17]. The disease is now better understood and, since 1960, no fatalities have been reported. Cases may have gone unreported, however, since it is not required that this disease be reported to public health officials [18].

Deicing Salt. In areas where winter snows are heavy and the use of salt for deicing is correspondingly large, chloride and sodium in drinking water have risen alarmingly [19]. This is especially true where drinking water sources are close to highways. In Massachusetts, the town of Burlington passed a salt ban on all town-maintained roads after chloride in one source of the town's drinking water reached 272 ppm—32 ppm in excess of the drinking water standard. Weston, another Massachusetts town, has been unable to use its public water supply since May 1970 and has had to import water from the Boston metropolitan system. The problem is general in Massachusetts and presents the greatest health danger to people suffering from heart disease, hypertension, and other conditions requiring low salt intake such as dropsy, renal disease, cirrhosis of the liver, obesity, and several pregnancy complications. The American Heart Association (AHA) recommends a limit of 22 ppm of sodium (a concentration reached at 59 ppm of chloride) for people whose diets are restricted to less than a quarter of the normal adult intake of four grams of sodium per day. The AHA warned that the drinking water in 62 communities in Massachusetts contained excess amounts of sodium in 1970 and, in the same year, the state's Commissioner of Public Health alerted physicians to the problem.

The problem is by no means confined to Massachusetts, however. Chloride concentrations have been measured in highway runoff at 10,250 ppm in Wisconsin and 1900–4500 ppm in Chicago, jumping to 11,000–25,000 when salt was applied to roads during storms [20]. If such heavily contaminated runoff continues, the problem which surfaced first in Massachusetts will eventually reach other communities.

In addition to the actual deicing agents sodium chloride and calcium chloride, other chemicals are associated with road deicing: Ferric ferrocyanide and sodium ferrocyanide are sometimes added to eliminate caking of the salt in storage and sodium hexametaphosphate and chromate salts to inhibit au-

When the snow melts, the salt remains. Some Massachusetts towns have had their drinking water so heavily polluted with deicing salt that it has become undrinkable.

Passing cars have covered these sumac branches with salt spray from deiced roads.

tomobile corrosion. In the presence of sunlight, for example, sodium ferrocyanide liberates cyanide—a highly toxic chemical to humans and wildlife [21].

Chlorinated Hydrocarbons. A less obvious and more insidious threat to human health arises from such long-lived chemicals as the chlorinated hydrocarbon pesticides such as DDT, which are transported to all parts of the globe in the air and accumulate in the environment. Early toxicity studies, which showed that healthy adults could ingest as much as 0.5 milligram per kilogram of body weight (mg/kg) per day of DDT (or 35 mg per person) in some cases for as long as 21 months without any adverse effects [22], were for many years accepted as reassurance that these pesticides posed no human health problems—in spite of the fact that people in various countries of the world where measurements have been made have stored from 9–15 ppm of DDT in their bodies [23]. As evidence of harm from DDT to both wildlife and experimental animals has begun to accumulate, however, concern has gradually shifted from immediate toxicity to more indirect effects [24].

Infant mortality in the offspring of female rats was found to rise with increasing amounts of DDT fed to the mothers and passed on to the newborn babies in the mothers' milk. Growth and activity in the surviving offspring were also adversely affected. When the mother rats were on a protein-deficient diet, the effects of the pesticide were magnified [25]. Although the amounts of DDT used in the experiment were higher than those to which humans are exposed, it has been found that DDT concentrates in the milk of human mothers and is passed on to their babies, too. In some Asian and African countries, infants are nursed for two to three years. In these same countries, DDT exposure may be particularly high and protein in the mothers' diets particularly low.

Other evidence is accumulating that DDT does have harmful effects on man and animals. One of the principal physiological effects of DDT in wild birds is its interference with reproduction by altering shell thickness. DDT also affects the enzymatic activity of the liver in the rat, according to reports by Michael T. Koeferl and R. E. Larson of Oregon State University [26] and Florence K. Kinoshita of the University of Chicago [27], thus reducing one of the body's important defenses against ingested poisons. Chlorinated hydrocarbons may be quite harmless while they remain stored in fat, but weight loss which mobilizes the fat may release them to other parts of the body where they are not so harmless. Serious health effects were found in mice after long-term exposure to DDT or dieldrin (another of the chlorinated hydrocarbons) was followed by a stress situation in which their body fat was mobilized [28]. Relatively large amounts of DDT and DDE (a compound produced when DDT breaks down) are transferred across the placental barrier from human mothers to their unborn children [29].

All these experimental findings are provocative, but not conclusive proof of harm to people. As with most environmental health problems, the thread of cause and effect is difficult to unravel. Research in the field is still rudimentary, even as DDT contamination continues to build up. Although the use of

An adult peregrine falcon eating one of its own eggs—a symptom that has been observed in predaceous bird populations experiencing reproductive failure. The widespread use of chlorinated hydrocarbon compounds has been implicated in the reproductive failure of the peregrine falcon in England and on the East coast of the United States.

DDT is now restricted in the United States, large amounts are still manufactured here and exported to Asian, African, and South American countries. Furthermore, the persistence of these compounds and their global circulation means that they would be present throughout the human environment for a long time even if their use were halted tomorrow.

The polychlorinated biphenyls (PCBs) are similar to DDT in their action: They are persistent in the environment, insoluble in water, and soluble in animal fat. PCBs have been incorporated into many industrial products, such as plastics, rubber, electrical products, and brake linings.

PCBs were recognized as an occupational hazard in the early 1930s. Ten years later, after repeated outbreaks of chloracne (a skin disease) and cases of digestive disturbances and even death from atrophy of the liver among exposed workers, a PHS authority warned that PCBs should not be allowed to come into contact with people [30]. Nevertheless, the chemicals were allowed to escape into the environment and were discovered in fish flesh in

1966 by Swedish scientists. Evidence now shows that PCBs and DDT are globally distributed [31]. It is estimated, for example, that the Great Lakes contain 100 tons of PCBs [32]. The PCBs may be ingested by fish that are then eaten by humans, or fish containing PCBs may be used for making animal feed and the animals fed may, in turn, be eaten by humans.

Some of the toxic effects of PCBs on humans were studied in Japan after an epidemic of PCB poisoning from contaminated rice oil occurred there in 1968 [33]. At last count in 1971, 1057 people were believed to have been affected. The disease, called Yusho (oil disease), caused eye discharge and a swelling of the upper eyelids, followed by acne, changes in skin pigmentation, and a feeling of weakness and fatigue. Even babies born to mothers who had eaten the contaminated oil showed some of the symptoms, especially darkening of the skin and nails and increased eye discharge. Three years after the incident an examination of the patients showed that the condition had improved in some, but had worsened in others.

Since PCBs are widely distributed in the environment, it is now important to know whether like DDT, they can be found in humans. A Human Monitoring Survey of pesticide levels in the U.S. population initiated in 1967 included a study of the levels of PCBs in the fat tissue of humans. PCBs are widespread in the population according to the preliminary data collected thus far. Of 637 samples analyzed, 31 percent contained at least one ppm of PCBs, 20 percent contained less than one ppm of PCBs, and the remainder were negative. Positive samples were received from every hospital in every city and state surveyed [34].

Oil. Oil is another persistent chemical contaminant with profound biological effects. One of the most complicated natural mixtures on earth, petroleum contains a wide variety of hydrocarbons, including some that are toxic and some that are carcinogenic. A year after an oil spill near the Woods Hole Oceanographic Institute in Massachusetts, some of these hydrocarbons could still be found in the coastal sediments. They became concentrated in the fatty parts of marine organisms such as bottom invertebrates and can be passed on up the food chain until they eventually reach man [35]. Oil may also concentrate other fat-soluble poisons present in the water such as insecticides, making these pollutants available to marine food chains in higher concentrations than would otherwise be the case [36].

Chlorine. As sources of unpolluted water become fewer and as the quantities of sewage dumped into specific waterways increase, disinfection requires ever larger amounts of chlorine. Another factor responsible for increasing the use of chlorine is that some chemical pollutants—ammonia, cyanide, and various organics, for example—combine with chlorine, reducing its effectiveness as a bacteriocide. More and more communities therefore are finding that their drinking water frequently tastes or smells of chlorine. Chicago, which has increased its use of chlorine by 67 percent over the past 20 years (as noted in "Assault on a Lake") is by no means alone.

But the problem of the interaction of chlorine and other chemicals is more serious than the taste or smell of chlorine. The entire range of possible reactions that might occur between chlorine and other water contaminants has not been studied. A particular danger, for example, would be the formation of chlorinated hydrocarbons. Included in the mixture of chemicals found in the Lower Mississippi survey were biphenyls and a number of benzene compounds. In the presence of iron, the biphenyls may react to form PCBs. In the laboratory, it is very easy to produce gamma-benzenehexachloride by bringing chlorine and benzene together in the presence of ultraviolet light. Two of the commercial names for this compound are Gammexane and Lindane, both insecticides. The possibility that this insecticide is being produced in the river should be investigated [37].

The problems resulting from the use of chlorine (as well as chlorine's possible ineffectiveness in killing viruses, to be discussed later) make it desirable to find an alternative water purifier. The hazard of transporting chlorine is a further incentive to seek a safe and effective substitute. Five million tons of chlorine were transported in 1967 [38], and the amount probably continues to increase. In truck or railroad accidents involving chlorine, the chemical may vaporize and expose people in the vicinity to highly toxic chlorine gas—one of the first gases used in warfare—or liquid chlorine accidentally dumped on land or in water may kill all soil or aquatic life in the vicinity of the accident.

Ozone, which has been used extensively in Europe as a disinfectant, is now being considered by some sanitary engineers in the United States as a possible alternative to chlorine. It has been claimed that ozone is more effective than chlorine in killing viruses and that ozone would eliminate some of the problems created by chlorine's reactivity with organic chemicals [39] as well as some organic substances responsible for objectionable colors, tastes, and odors in drinking water.

Both chlorine and ozone may be hazardous to workers who manufacture and handle them. Ozone is explosive in the gas, liquid, and solid states, and thus presents an additional hazard. On the other hand, ozone's explosiveness would require that it not be transported, but prepared at the site of disinfection, thus eliminating any transportation hazard. At present, the cost of ozone treatment is higher than chlorine—according to one case study, three times as high [40]. However, since much less chlorine is needed to obtain disinfecting power, and new production systems for ozone are being designed, it is expected that the costs can be substantially reduced. There are obviously many factors to be considered in comparing their relative safety and effectiveness, and their respective cost in monetary and environmental terms.

Biological Hazards

As mentioned earlier, the standard used to measure the biological hazard to humans is the coliform count, not because coliform bacteria are themselves disease organisms, but because they are useful indicators of the presence of disease organisms. Coliform bacteria live in fecal matter in human intestines,

History, ecology, and music are part of the freight carried by the sloop *Clearwater,* a replica of the sloops that used to ply the Hudson in the days when its waters were clear. Hudson River Sloop Restoration, Inc., intends to keep sailing until those waters are clear again and is bringing that day closer with an imaginative combination of information about river pollution that includes analysis of water samples, action against polluters, and education of the people in riverside communities. The *Clearwater,* all wrapped up with sails and songs (the latter from folk singer Pete Seeger and other lesser-known but equally dedicated musicians), has carried its message to Long Island and Connecticut, and has inspired people in Chesapeake Bay, the Great Lakes, and San Francisco to plan the construction of crafts similarly linked to the past and the future of their own waterways.

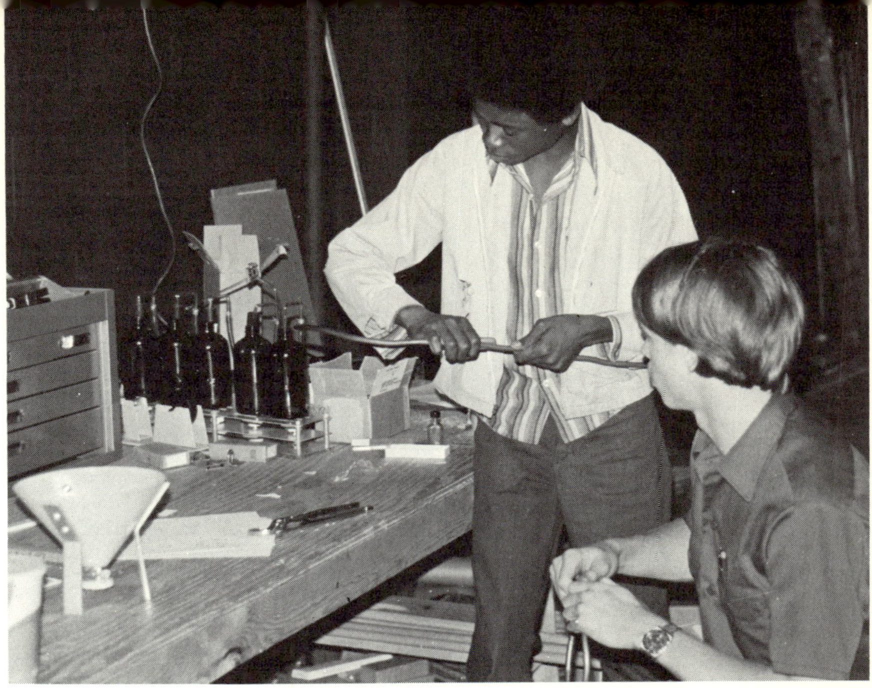

In the Institute of Environmental Education Summer Study Program, students and teachers examined a whole range of water pollution problems. Above, some equipment devised and built by students; below, students count bacteria colonies to develop a profile of stream pollution.

while they may be found in water in the absence of feces, excessive numbers of coliforms are strong evidence that the water is contaminated with bacteria from human feces and that microorganisms of human disease are also likely to be present. Coliform bacteria are easy to recognize and to count, while the identification of many separate bacteria in the excretions of people suffering from various diseases would be extremely difficult to isolate and to identify.

Waterborne Viral Diseases. Since viral diseases can also be disseminated in the same way, it has been assumed that water free of contamination by intestinal bacteria—as measured by the coliform count—would be unlikely to be dangerously contaminated with viruses. The ratio of viruses to coliforms in sewage is about 15 to a million [41].

However, in many cases, fewer organisms are needed to cause a viral than a bacterial infection. For example, in one experiment it was found that 30 percent of human volunteers receiving *one* polio virus became infected [42], while a single bacterium is rarely, if ever sufficient to infect. Also, under some circumstances, viruses can survive in water for longer periods of time than bacteria.

A virus becomes an important agent in transmitting disease through a water supply when it is excreted in urine or feces, is able to survive in water, and is capable of infecting people when ingested. A number of *enteroviruses* fit this description. Although all enteroviruses can multiply in the human intestinal tract, their effects are not limited to that part of the body. They include polio viruses which may attack the central nervous system resulting in paralysis and the coxsackie viruses which may produce a disease similar to polio and also may produce congenital heart abnormalities in infants. (This occurs when the uterus of the mother has been infected with the virus.) The enteroviruses also include influenza-causing viruses and infectious hepatitis. Approximately 100 viruses may be included in the category of viruses that could transmit disease through a water supply. Except for infectious hepatitis, all of these viruses have been isolated and identified and have been found in human sewage [43].

Evidence for waterborne infectious hepatitis was given in a 1966 report in the *American Journal of Public Health* [44]. Infectious hepatitis is a prevalent viral disease—for example 72,651 cases were reported in 1961. In the journal report, 48 waterborne outbreaks of infectious hepatitis were analyzed. Of these, 14 outbreaks (29,381 cases) were associated with public water supplies and 34 outbreaks (1976 cases) with private water supplies. It should be noted that many of these water supplies had not been chlorinated. The largest recent outbreak of infectious hepatitis occurred from December 1955 to January 1956 in New Delhi, India. There the public water source became contaminated with raw sewage. Although the water was subsequently filtered and chlorinated, an epidemic—approximately 28,000 cases—of infectious hepatitis occurred. There was no increase in typhoid fever or bacterial intestinal diseases during the hepatitis epidemics, however, indicating to water hygiene experts that viruses may be more resistant to chlorine treatment than bacterial disease agents.

There were 142 outbreaks of waterborne diseases—gastroenteritis and diarrheal disease—affecting 18,000 people reported between 1946 and 1960 [45]. These cases were believed to be by viruses; bacteria are not often found to be the causative agent.

A small number of polio outbreaks can possibly be attributed to contaminated water supply systems. In November of 1953, for example, an outbreak of polio occurred in Edmonton in the province of Alberta, Canada. Cases were uniformly distributed throughout the city and no common sources of food, milk, or public contact could be established; an additional factor ruling out personal contact was that the cases occurred too closely together to allow enough time for the incubation period. Cases of polio had occurred in Devon, a town upstream from Edmonton, earlier in the season. At this time, the chlorination of Devon's sewage effluent had been inconsistent and polio viruses could have been carried to the Edmonton water treatment plant. Although the Edmonton plant then settled, softened, filtered, and chlorinated the water, this was thought to be the most likely route for the polio virus to have taken [46].

More recently, a polio-like virus was found in drinking water which had been tested as safe. According to a report by two researchers at the University of Michigan and a health official from the Michigan Department of Health, this was the first time a virus had been isolated from drinking water [47]. On January 13 and 20 of 1970, complaints of intestinal illness, with symptoms including nausea, vomiting, and diarrhea, were made to Michigan health officials. Since all of the people affected had eaten at the same restaurant, health officials tested the food there, but no bacteria that could have caused intestinal disturbance were uncovered.

Subsequently, water samples were tested. The restaurant drew water from a 100-foot well on the edge of a drainfield which received waste water; thus the well could easily have become contaminated. Using routine procedures for bacteriological testing, no *Escherichia coli*, *Salmonella*, or *Shigella* were found. The examiners then concentrated the water samples, making the chances of detecting bacteria more likely. When this was done, both bacteria and a polio-type virus were recovered. Since no attempt was made to isolate the virus from the patients, there is no direct proof that it actually caused the intestinal illness. However, the study does show that viruses can remain alive in water and that routine bacteriological surveys may not be adequate protection against disease.

A Lack of Knowledge. Questions concerning the adequacy of PHS bacteriological standards and the treatment processes designed to safeguard them were raised at hearings on "Safe Drinking Water" held in May 1971 by the Subcommittee on Public Health and Environment of the Committee on Interstate and Foreign Commerce of the House of Representatives [48]. The hearings were prompted in part by several major disease outbreaks in recent years: Included were a large outbreak of gastroenteritis (thought to be viral in origin) in 1968 in Angola, New York, and an outbreak of infectious hepatitis affecting the Holy Cross football team in 1969.

One of the problems raised at the hearings was the general lack of knowledge of the spread of viral infection by water. It was pointed out that while there is the prospect of increased reuse of water, there have been few studies of viruses in reuse systems over long periods of time and we know little about the movement of viruses in groundwater or the ability of water treatment plants to remove viruses [49]. The 1962 PHS drinking water standards do not include limits for viruses. Although the Advisory Committee on Revision of U. S. Public Health Service 1946 Drinking Water Standards discussed setting standards for viruses, it was concluded that not enough was known about them to limit them or to recommend a routine procedure for their examination in water supplies. However, it was recommended that those laboratories equipped to isolate and identify viruses should evaluate enteroviruses in treated water [50].

A scientific paper submitted to the "Safe Drinking Water" hearings summarized the results of a variety of reports on the efficiency of various aspects of water treatment in deactivating viral contaminants [51]. There are few field data, but laboratory experiments revealed a number of deficiencies. The storage of water—one step in some water treatment systems—does not always help to reduce virus count. In warm months, it may take several weeks to obtain a 99.9 percent reduction of viruses; during prolonged cold weather, viruses may remain active for months. There is also evidence that infectious hepatitis agents survived ten weeks in well water stored in a laboratory. Filtering—another step in water treatment—may be ineffective in removing viruses if the flow of water is too rapid [52]. For those plants with a multistep system, the last step in the treatment process is usually chlorination. Some laboratory data show that many viruses are more resistant than bacteria to chlorine treatment. For example, a 99.9 percent removal of *E. coli* was obtained in five minutes with a chlorine concentration of 0.04 ppm, while it took 30 minutes to remove 99 percent of the infectivity of one type of polio virus with a chlorine concentration of 0.11–0.20 ppm [53].

When chlorine is added to water, it reacts initially with the chemicals in the water. The chlorine remaining after these initial reactions—called the *free residual chlorine*—is what is actually available for killing bacteria. The PHS recommends that a free chlorine residual of 0.2 ppm be obtained at the treatment plant, and a residual of 0.05–0.1 ppm be maintained in the distribution system [54]. Most states do not set standards for chlorine residuals, but laboratory studies have indicated that a chlorine residual higher than these recommended concentrations would be necessary to kill viruses. In a multistep treatment process, virus removal could be increased by chlorinating twice—at the beginning of the process and again at the end [55].

Flocculation or *coagulation*—the clumping of particles so that they will settle which is one part of the clarification step in water treatment—can be very effective in removing viruses if pH, coagulant concentration, and mixing are carefully controlled. Laboratory data show that a 99 percent virus removal is possible in a good flocculation system [56]. Unfortunately, in the community water supply study discussed at the beginning of this chapter, only 17 percent of the systems surveyed included clarification as part of their water treatment, and half the systems neither clarified nor disinfected the water [57].

O. C. Liu of the EPA's Northeastern Water Hygiene Laboratory, in a paper submitted to the "Safe Drinking Water" hearings, summarized the status of knowledge about the efficiency of present water treatment processes by saying that, although some of the water treatment processes now in use are relatively efficient in removing or destroying enteroviruses, only a small number of viruses have been studied and a general pattern for all types and groups of the human enteroviruses is not yet clear [58]. Liu added that little is known about the virus density of various surface waters and thus "it is impossible to accurately assess what safety margin the present treatments may provide when polluted sources are used." [59] In his opinion, there is always a potential danger of waterborne hepatitis outbreaks, and they will recur if the conditions of untreated water or the methods used to treat water favor the increase of the virus population or if, as may be the case, the transmission of low virus concentrations by water is responsible for some portion of the present incidence of infectious hepatitis [60].

Soil Bacteria. Rising bacterial counts in some waterways may not be the result of increased intestinal bacteria, but an indication of the presence of soil bacteria. Soil microorganisms—fungi, bacteria, protozoa—may be *pathogenic* (capable of producing disease in people). Although these soil microorganisms may be washed out of the soil and into water frequently, they usually represent no hazard to people because, under natural conditions, they cannot live and reproduce in water. Barry Commoner has described surface waters, therefore, as a "very effective biological barrier to the movement of pathogenic microorganisms from the soil to man because the water ordinarily contains insufficient organic matter to support the growth of the pathogen." [61] But as eutrophication and organic pollution increase, the increased organic content of water gives it, like soil, the ability to support heavy bacterial growth. This, in turn, can lead to the growth of some types of highly toxic amebas.

Medical journals have recently reported cases of amebic meningoencephalitis, a disease caused by a free-living soil ameba. The infectious stage of this organism is stimulated by the presence of bacteria often found in abundance in ponds and lakes toward the end of summer as well as near sewage outfalls and wherever there is an adequate source of organic material that will serve as food for bacteria. A recent report described three fatal cases of amebic meningoencephalitis in Florida; the victims had been swimming in small, land-locked lakes, and the amebas recovered from their spinal fluid were also found in lake water samples [62].

When soil washes into the water, it carries with it some of the many pathogenic microorganisms that inhabit it, but these microorganisms find water an inhospitable environment because it usually does not contain enough organic matter to support their growth. Swimmers are therefore protected from diseases to which they would otherwise be exposed. But if a pond or lake becomes eutrophic or heavily polluted with organic matter, it may acquire the ability to support the growth of some of these microorganisms and the protective barrier may break down.

Soil molds recently isolated from water heavily polluted with organic matter included some which can cause infections if they become established in the human body and others which produce an active carcinogen [63].

Thus water—the sustainer and protector of life—can become a transmitter of disease organisms and of both organic and inorganic chemical contaminants to man. It has been estimated that each year 500 new chemicals are added to the list of industrial chemicals in use. It becomes questionable whether those responsible for monitoring and setting water standards, as well as devising methods for water treatment, can catch up with the ever-increasing chemical load. Current standards and treatment processes are not adequate to provide water pure enough for drinking and swimming as long as the sources of community water supplies and the sites of recreational water receive domestic waste, agricultural and construction runoff, and industrial effluent. Reevaluation of present laws and institutions dealing with water is necessary for the protection of human health.

[1] Leland J. McMcCabe, James M. Symons, Roger D. Lee, and Gordon G. Robeck, "Survey of Community Water Supply Systems," *Journal of the American Water Works Association*, November 1970, Vol. 62, No. 11, 670–87.

[2] The presence of the coliform group of bacteria (a group which includes *Escherichia coli*, an inhabitant of the human intestine) in a drinking water supply indicates that it is contaminated with fecal matter. The Public Health Service requires that a minimum amount of water samples be routinely checked to be sure that the number of coliform bacteria is within the limits set in PHS drinking water standards.

[3] McCabe et al., "Survey of Community Water Supply Systems," adapted from Tables 5 and 14, pp. 674 and 679 (see [1]).

[4] W. H. Durum, J. D. Hem, and S. G. Heidel, "Reconnaissance of Selected Minor Elements in Surface Waters of the United States, October 1970," *Geological Survey Circular 643*, (Washington, D. C.: U. S. Government Printing Office, 1971).

[5] *Industrial Pollution of the Lower Mississippi River in Louisiana*, Environmental Protection Agency, Surveillance and Analysis Division, Region VI, Dallas, Texas, April 1972, pp. 1–146.

[6] These discharge values did not include a correction for contamination already in the water at the intake, but discharges at other plants where such allowances were made included values up to 889 pounds for lead, 38 pounds for cadmium, and 188 pounds for chromium.

[7] Paul R. Craig and Edward Berlin, "The Air of Poverty," *Environment*, June 1971, Vol. 13, No. 5, 2–9.

[8] Julian McCaull, "Building a Shorter Life," *Environment*, September 1971, Vol. 13, No. 7, foreword, p. 2.

[9] *Drinking Water Standards, 1962,* Public Health Service, U. S. Department of Health, Education and Welfare, Washington, D. C., p. 51.

[10] *Industrial Pollution of the Lower Mississippi River,* adapted from Tables 17, 18, and 19, pp. 75–85 (see [5]).

[11] Ibid., pp. 67 and 69.

[12] "Organic Chemical Pollution of Fresh Water," *Water Quality Criteria Data Book, Volume 1,* Environmental Protection Agency, Water Quality Office, December 1970.

[13] Sheldon O. Smith and Joseph H. Baier, "Report on Nitrate Pollution of Groundwater," Nassau County Department of Health, Mineola, New York, December 1969.

[14] "Nitrates in Ground Waters of the Central Coast Area," The Resources Agency, State of California, Department of Water Resources, San Joaquin District, June 1971.

[15] "Nitrate Sources in Surface Waters," CBNS Notes, Center for the Biology of Natural Systems, Washington University, St. Louis, Missouri, September–October 1970, Vol. 3, 4.

[16] "Poisoning the Wells" *Environment* (staff report), January–February 1969, Vol. 11, No. 1, 16–23, 45.

[17] Douglas H. K. Lee, "Nitrates, Nitrites, and Methemoglobinemia," *Environmental Research,* Vol. 3, 484–511.

[18] Lois Ann Shearer, John R. Goldsmith, Clarence Young, Owen A. Kearns, and Benjamin R. Tamplin, "Methemoglobin Levels in Infants in an Area with High Nitrate Water Supply," *American Journal of Public Health,* September 1972, Vol. 62, No. 9, 1174.

[19] H. Hugh McConnell and Jennifer Lewis, ". . . Add Salt to Taste," *Environment,* November 1972, Vol. 14, No. 9, 38–45.

[20] R. E. Hanes, Zelazney, and R. E. Blaser, "Effects of Deicing Salts on Water Quality and Biota," *National Cooperative Highway Research Program Report No. 91,* Highway Research Board, Washington, D. C., 1970; F. H. Schraulnagel, "Pollution Aspects Associated with Chemical Deicing," *Highway Research Record Report No. 193 on Environmental Considerations in the Use of Deicing Chemicals,* Highway Research Board, Washington, D. C., 1967; and R. H. Sullivan, "Effects of Winter Storm Runoff on Vegetation and as a Factor in Stream Pollution," paper presented at the American Public Works Association, 7th Annual Snow Conference, Milwaukee, Wisconsin, April 12, 1967, cited in McConnell and Lewis, ". . . Add Salt to Taste," p. 39 (see [19]).

[21] *Environmental Impact of Highway Deicing,* Environmental Protection Agency, Edison Water Quality Laboratory, June 1971, pp. 65–70.

[22] Wayland S. Hayes, Jr., *Clinical Handbook on Economic Poisons,* revised edition, Public Health Service Communicable Disease Center, Toxicology Section, Atlanta, Georgia, 1963, p. 59.

[23] Julian McCaull, "Questions for an Old Friend," *Environment,* July–August 1971, Vol. 13, No. 6, 6.

[24] Ibid., pp. 2–9.

[25] Ibid., pp. 2–3. Report on a paper by Dr. M. S. Fahim presented at the 55th Annual Meeting of the Federation of American Societies for Experimental Biology in Chicago, Illinois, April 12–17, 1971.

[26] Ibid., p. 6. Report on a paper by Drs. Michael T. Koeferl and R. E. Larson presented at the 55th Annual Meeting of the Federation of American Societies for Experimental Biology (see [25]).

[27] Ibid., pp. 6–7. Report on a paper by Dr. Florence K. Kinoshita presented at the 55th Annual Meeting of the Federation of American Societies for Experimental Biology.

[28] Ibid., pp. 7–8. Report on a paper by G. L. Paulson presented at the 55th Annual Meeting of the Federation of American Societies for Experimental Biology.

[29] J. A. O'Leary, J. E. Davies, W. F. Edmundson, and G. A. Reich, "Transplacental Passage of Pesticides," *American Journal of Obstetrics and Gynecology,* March 15, 1970, Vol. 106, No. 6, 939.

[30] Leonard Greenburg, "Chlorinated Naphthalenes and Diphenyls," *Industrial Medicine,* August 1943, Vol. 12, 520–22.

[31] Robert Risebrough, with Virginia Brodine, "More Letters in the Wind," *Environment,* January–February 1970, Vol. 12, No. 1, 16–27.

[32] Ian C. T. Nisbet and Odel F. Sarofim, "Rates and Routes of Transport of PCBs in the Environment," *Environmental Health Perspectives,* April 1972, Experimental Issue No. 1, 32.

[33] Masanori Kuratsune, Takesumi Yoshimura, Junichi Matsuzaka, and Atsuko Yamaguchi, "Epidemiologic Study on Yusho, a Poisoning Caused by Ingestion of Rice Oil Contaminated with a Commercial Brand of Polychlorinated Biphenyls," *Environmental Health Perspectives,* April 1972, Experimental Issue No. 1, 119–28. Reports given at a conference on PCBs sponsored by the National Institute of Environmental Health Sciences, Rougemont, North Carolina, December 20–21, 1971.

[34] Anne R. Yobs, "Levels of Polychlorinated Biphenyls in Adipose Tissue of the General Population of the Nation," *Environmental Health Perspectives,* April 1972, Experimental Issue No. 1, 79–81.

[35] M. Blumer, H. L. Sanders, J. F. Grassle, and G. R. Hampson, "A Small Oil Spill," *Environment,* March 1971, Vol. 13, No. 2, 2–12.

[36] R. Hartung and G. W. Klinger, "Concentrations of DDT by Sedimented Polluting Oils," *Environmental Science and Technology,* 1970, Vol. 4, 407, cited in Blumer et al., "A Small Oil Spill," p. 12 (see [35]).

[37] Suggested by Dr. Charles Lee, Center for the Biology of Natural Systems, Washington University, St. Louis, Missouri.

[38] Eric Albone and Julian McCaull, "Freighted with Hazard," *Environment*, December 1970, Vol. 12, No. 10, 21.

[39] K. M. Yao, "Is Chlorine the Only Answer?" *Water and Wastes Engineering*, January 1972, Vol. 9, No. 1, 30–31.

[40] Ibid.

[41] N. A. Clarke, G. Berg, P. W. Kabler, and S. L. Chang, "Human Enteric Viruses in Water: Source, Survival, and Removability," in *Advances in Water Pollution Research*, Volume 2, Proceedings of the International Conference, London, 1962 (London: Pergamon Press, 1964).

[42] *Safe Drinking Water, Hearings before the Subcommittee on Public Health and Environment of the Committee on Interstate and Foreign Commerce*, House of Representatives, May 24, 25, and 26, 1971. Cited in testimony submitted for the record by William D. Ruckelshaus, Administrator, Environmental Protection Agency, and James McDermott, Director, Division of Water Hygiene, Environmental Protection Agency, p. 379.

[43] O. C. Lui, "Effect of Chlorination on Human Enteric Viruses in Partially Treated Water from Potomac Estuary," Environmental Protection Agency progress report, Northeastern Water Hygiene Laboratory, November 1970, a paper submitted to *Safe Drinking Water*, p. 398 (see [42]).

[44] Floyd B. Taylor, James H. Eagen, H. F. C. Smith, and Ronald F. Coene, "The Case for Waterborne Infectious Hepatitis," *American Journal of Public Health*, December 1966, Vol. 56, No. 12, 2093–105.

[45] S. R. Weibel, F. R. Dixon, R. B. Weidner, and L. J. McCabe, "Waterborne Disease Outbreaks, 1946–1960," *Journal of American Water Works Association*, September 1964, Vol. 56, Pt. 2, No. 8, 947–58.

[46] G. M. Little, "Poliomyelitis and Water Supply," *Canadian Journal of Public Health*, March 1954, Vol. 45, 100–102.

[47] Walter N. Mack, Uue-Shoung Lu, and Donald Coohon, "Isolation of Poliomyelitis Virus from a Contaminated Well," *Health Services Reports*, March 1972, Vol. 87, No. 3, 271–74.

[48] *Safe Drinking Water*, p. 1 (see [42]).

[49] Charles C. Johnson, Jr., statement submitted to *Safe Drinking Water* on behalf of the American Public Health Association, p. 167 (see [42]).

[50] *Drinking Water Standards, 1962*, p. 19 (see [9]).

[51] Lui, "Effects of Chlorination on Human Enteric Viruses," pp. 402–405 (see [43]).

[52] Ibid., p. 403.

[53] Ibid., data from Table 7, p. 427.

[54] Edmund J. Raubusch, "Water Disinfection Practices in the United States," *Journal of the American Water Works Association,* November 1960, Vol. 52, No. 11, 1416–426.

[55] Norman A. Clarke and Sluh Lu Chang, "Enteric Viruses in Water," *Journal of the American Water Works Association,* October 1959, Vol. 51, No. 10, 1310–311.

[56] Lui, "Effect of Chlorination on Human Enteric Viruses," p. 403 (see [43]).

[57] McCabe et al., "Survey of Community Water Supply Systems," data from Table 3, p. 672 (see [1]).

[58] Lui, "Effect of Chlorination on Human Enteric Viruses," pp. 405–406 (see [43]).

[59] Ibid., p. 406.

[60] Ibid., pp. 405–406.

[61] Barry Commoner, *The Closing Circle* (New York: Alfred A. Knopf, 1971) p. 221.

[62] Cecil G. Butt, "Primary Amebic Meningoencephalitis," The *New England Journal of Medicine,* June 30, 1966, Vol. 274, No. 26, 1473–76.

[63] Commoner, p. 233 (see [61]).

The Limits of Patchwork

On a hot day not long ago, children and their parents stood in line to swim in a shallow artificial basin at Santee, in southern California. With the Pacific Ocean less than one hour away, the lure of the man-made swimming hole was not immediately apparent. Their attraction was a mixture of curiosity and some pride in a community experiment: The basin contained sewage water that had been highly treated or *reclaimed* to make it safe enough for bathing [1].

As of 1971, some 45 California sewage reclamation systems were producing water for ornamental and recreational lakes and for the irrigation of golf courses, parks, and freeway landscapes. Sewage wastewater purified by advanced techniques is used for industrial processes, primarily cooling, in some 25 U.S. cities [2]. Studies in Denver and Dallas are being conducted to determine whether discharges from secondary treatment plants can be retreated with advanced methods to make the water pure enough to drink [3]. In 1971, the American Water Works Association (AWWA) gave official support to investigating the use of reclaimed municipal waste water for industry, irrigation, and recreation, and, where appropriate, for renewal of underground water supplies [4]. Abroad, Windhoek, the capital of Southwest Africa, in 1968 became the first city in the world to reclaim municipal treatment plant effluent on a large scale for drinking water: The treated waste water made up about one-third of the city's total domestic water supply [5].

In reviewing some of these trends, one author in the field of waste management commented:

> One should realize there is no such thing as new water; nearly all water on the earth's surface has been used and reused in one way or another. Perhaps one

> should refer to the water that carries our wastes as "transport water." Then in treatment plants the pollutants can be removed from the water to end up with plain water. The result might be far more acceptable to the public than trying to get them to drink "reclaimed sewage." [6]

Recognition that waste-bearing water is a means of transporting rather than "getting rid of" wastes might also encourage a healthy reappraisal of the notion that the nearest stream is a good waste disposal site. We have, in effect, created the need for a science of water pollution by first creating disposal technology based on the supposed capacity of water to assimilate wastes. But now the science of water pollution, like an ungrateful child, argues loudly that the parent technology of water disposal is dangerously old fashioned and should be curbed, if not forcibly retired.

Water Conservation

The degree to which such retirement is realistic depends upon two major factors. One is the extent to which water scarcity forces municipalities and industries to reject conventional waste disposal practices in favor of water conservation. The preceding examples of wastewater reuse demonstrate this influence, particularly in the Southwest and West where fresh water is limited. The second major factor is the amount of pressure for clean water generated by the public and subsequently reflected in pollution-control regulations. As the regulations tighten, the incentive for water reuse grows, particularly in industry. For example, the release of ammonium nitrate, a chemical fertilizer, from the Farmers Chemical Association plant in Harrison, Tennessee, prompted criticism from the Tennessee Stream Pollution Control Board in 1968. The Board charged that nitrogenous materials from the plant contaminated the domestic water supply for Chattanooga, a short distance downstream on the Tennessee River. This pressure moved the officials of the cooperative plant to introduce a closed-cycle wastewater system—the first in the chemical fertilizer industry [7]. Plant waste water was collected in large settling ponds, filtered to remove undissolved solid waste, and then put in contact with special resins to remove the ammonium and the nitrate ions separately by a process called *ion exchange.* The resins were later treated chemically to release the ions, and this allowed recovery of about six tons of ammonium nitrate daily. The salvaged ammonium nitrate solution was then blended to formulate the plant's final product—a liquid mixture of ammonium nitrate, urea, and water to be applied directly to soil.

Before the system was installed, an average of 900,000 gallons of waste water was discharged each day from the Tennessee plant. The conversion eliminated the discharge. Furthermore, rain water collected in the settling ponds compensated for water losses from evaporation or other causes and thus eliminated the need to tap the Tennessee River. In other words, use of water at the plant was internalized to reduce reliance on free-flowing water to a minimum; this proved financially beneficial to the plant in terms of recovered product and environmentally beneficial in terms of reduced water pollution.

A closed-cycle wastewater system appears to have limited application in the overall manufacture of ammonium nitrate, however. In 1971, only 31 of 58 U.S. producers made a solution as opposed to a solid product. The other 27 companies would have to dispose of the ammonium nitrate solution recovered from the ion exchange resins since the substance would still be in a waste rather than in a commercially usable form insofar as their particular manufacturing processes were concerned. It must also be remembered that nitrogenous fertilizer, when used in excessive amounts on the soil, may leak into and thereby contaminate surface waters.

Closed-cycle use of water in other industries is gaining favor, but total recycling is rare. Furthermore, reduced withdrawal of water as the result of recycling is offset by intensified industrial water use for increased production of goods [8]. At the same time, the trend in pollution control investment in industry has been toward purchase of equipment to treat waste water rather than toward revision of manufacturing processes to reduce pollution at the source. A survey of 789 companies in seven major industrial categories noted an increase of nearly 240 percent in annual capital expenditures for pollution control during 1965–1969 (a total of $33.2 million was spent in 1965, compared to $112.6 million in 1969). Analysis of the expenditures, however, showed that: "The greatest portion of this five-year expenditure program was devoted to the construction of wastewater treatment facilities, and only a small portion toward developing changes in the manufacturing process which would result in significant pollution reduction" [9].

Waste water laden with ammonium nitrate is purified in these ion-exchange units at the Farmers Chemical Association plant in Harrison, Tennessee. The ammonium nitrate is later recovered for commercial fertilizer.

Control Regulations

Continued pollution resulting from the slowness in conversion to less-polluting industrial processes is amplified by a technological trend in the other direction: The types of technology gaining favor in steelmaking, the canning of fruits and vegetables, chemical manufacturing, and farming actually increase either the amount of toxicity of water pollution, as we saw in "Thirsty Technology." Still another difficulty is that water pollution control regulations have emphasized waste treatment instead of changes in the productive processes themselves. At the national level in the United States, the Federal Water Pollution Control Act, prior to the 1972 amendments to be discussed shortly, required state-established, federally-approved water quality standards for the country's network of interstate lakes and rivers [10]. The legislation, as variously amended during 1956–1970, was a logical outcome of the historical development of water law. The law that evolved, however, was far from logical in terms of the interconnected nature of U.S. waterways. For example, waterways were classified according to intended purposes ranging from drinking water withdrawal to industrial consumption. The standards for water quality were then based on the levels of pollutants such as bacteria, toxic materials, and odor-producing substances that could be tolerated in the water according to the designated use. Thus, greater pollution was allowed in water classified for industrial use than in water designated for drinking purposes. The approach fails to protect the overall integrity of waterways since pollutants released in an area designated for industrial use can be easily carried into other areas classified as domestic or recreational.

This shortcoming was tacitly acknowledged in the 1972 amendments to the Federal Water Pollution Control Act, and passed into law by Congress on October 18, 1972, over the veto of the president [11]. Although the 1972 legislation retained numerous links with older waste control concepts, a number of departures were introduced which answered some of the criticisms leveled at former legislation by environmentalists and scientists. The 1972 law established a two-phase program with a national goal of eliminating the discharge of *all* pollutants into navigable waterways by 1985. Navigable waters include more than 90 percent of all natural bodies of water in the United States, so the legislation would, in effect, authorize nationwide water pollution controls in place of the present system of varying local and federal laws.

The first phase of progress toward the 1985 goal will require the introduction of the "best practicable pollution control technology" in existing industrial establishments by mid-1977. The best practicable pollution control technology will be determined for various categories of industry such as pulp and paper manufacturing, petroleum refining, iron and steel manufacturing, and food processing. The determination will be made by means of regulations established by the EPA to provide guidelines to identify, in terms of amounts of specific pollutants, the degree of effluent reduction attainable by current best practicable technology. Specific pollution controls needed to meet the so-called attainable levels will be determined locally on a plant-by-plant basis, but the national guidelines will specify the factors to be taken into account in making these decisions. These factors include the total cost of alternative technological controls as compared with the benefits in pollution reduction achieved, the age of the plant's equipment and facilities, the manufacturing processes involved, the engineering aspects of the control technology, and the environmental impact of pollution control equipment, such as high energy requirements. The guidelines will include the degree of pollution reduction obtainable by specific "best practicable" methods. The alternative methods will not be limited strictly to pollution treatment techniques, but will include changes in process and operating methods that can achieve the desired results. For example, the best practicable technology for a given plant might be a new operating procedure utilizing closed-cycle use of process water rather than external pollution controls.

EPA regulations also will establish comparable guidelines for the degree of effluent reduction attainable in existing industrial processes by the use of the best available as opposed to the best practicable technology. The former, required by mid-1983, will be the second phase in the move toward the 1985

This wastewater treatment facility at the GAF Corporation's industrial products plant in Joliet, Illinois, is said to have cost $2.6 million and to add $500,000 in plant operating expenses annually. Unfortunately, most industrial investment for water pollution control is devoted to treatment facilities, and only a small portion is applied to research and development of process changes which would result in the creation of less pollution.

goal of elimination of all pollutants. Overall, then, the new legislation is aimed at producing uniform effluent limitations for industries according to their specific categories and in light of various technological options. Similarly, a uniform first-phase goal of secondary treatment for municipal waste treatment plants is established for plants in operation by mid-1977. The second-phase, mid-1983 goal for these municipal plants requires more advanced treatment including, if practicable, the reclaiming and recycling of waste water. In approving federal grants for construction of waste treatment works, the EPA administrator is directed to encourage the development of systems that recycle sewage by using it to fertilize crops, that reclaim waste water, or that unrecycled pollutants do not allow to reenter the environment. The legislation authorizes surveys for reservoirs as a part of comprehensive water pollution control programs, but provides that "any such storage and water release shall not be provided as a substitute for adequate treatment or other methods of controlling waste at the source" [12]. In amplifying this provision, members of the joint Senate-House committee which prepared the final version of the legislation observed: "The Conference substitute specifically bans pollution dilution as an alternative to waste treatment" [13]. Reservoir water could, however, be used to augment stream flow to dilute general pollution from undefined sources such as farms and to foster recreation and navigation.

To ensure that industrial facilities comply with the new law, individual plants will be required to obtain licenses to construct or operate facilities that could discharge pollution into navigable waterways. Appropriate state agencies will issue the required permits to the plants, but the EPA will have ultimate authority over the permits or over the state programs under which they are issued. A side effect of this action will be the phasing out of the 1899 Refuse Act Permit Program, directed by the U.S. Army Corps of Engineers but stalled by court litigation initiated by both environmentalists and industrialists.

To provide for upgraded pollution control in industrial facilities constructed after enactment of the legislation, the EPA administrator is charged with establishing "standards of performance" for new facilities in at least 28 specified industrial categories. The standards will detail the degree of pollution reduction to be obtained by the best available demonstrated technology. Standards will take into account improved production processes and manufacturing techniques, in addition to conventional pollution treatment procedures.

Situations might arise in which one plant or, more likely, a group of plants discharges allowable amounts of individual pollutants, yet reduces water quality below levels specified by the standards. This might be due to particular stream conditions or to the combined impact of discharges from a cluster of factories. To guard against these problems, the older concept of water quality standards based on designated use is retained in the new law. In particular, more stringent effluent limits will be required if normal limits hinder the propagation of aquatic life and the maintenance of water recreation. Before these more stringent effluent limits could be set, however, hearings would be necessary to consider the balance between the relevant economic and social costs and benefits. In the broader context of state and regional plans for water pollution control, water quality standards will be used as a tool to identify waterways in which stricter than usual effluent limits will ultimately be needed according to a priority ranking that takes into account the severity of the pollution and the designated use of the water. Discharges of heated industrial water as well as other types of pollution must be assessed in this planning process, and plants generally will be required to use the best available technology to minimize the environmental impact of thermal discharges.

As a guide to factors that determine the impact of pollution in water, criteria will be published by the EPA which specify the effects of specific pollutants on water quality and on aquatic life, including effects on biological diversity, productivity, and stability. Guidelines also will be published on what happens to ocean life and to the productivity and stability of marine ecosys-

Air pollution? Yes. Vehicular exhaust has long been considered a source of air pollution, but it is an important source of water pollution as well—and one which escapes water pollution laws.

tems as the result of dumping wastes at sea. Human health, recreational values, and the persistence of ocean pollution also will be considered. Marine pollution guidelines will detail alternative disposal techniques, including land-based methods. Permits must be obtained for dumping at sea, even in the deep ocean, in accordance with information published in the guidelines.

The new legislation establishes much tighter control over direct pollution of the nation's waterways than did previous laws. It provides a good deal of flexibility in choosing alternative methods such as recycling and process change, rather than relying exclusively on waste treatment. This is an incentive to develop ecologically sound ways to dispose of or to reuse our wastes instead of automatically looking for the nearest stream in the hopes that it still has some dilution capacity left. The provision for permits for ocean dumping extends federal authority to most dumping activities and should restrict, although certainly not eliminate, this source of marine pollution. Finally, the national goal to eliminate the discharge of all pollutants into navigable waters by 1985 marks a welcome departure from the older concept that waterways were convenient natural waste disposal sites.

Controlled Pollution vs. No Pollution

Realization of the 1985 goal, however, will be a different matter. Despite formal acknowledgment of the widescale environmental implications of water pollution, the legislation is keyed to our technological capacity to control pollution rather than to a mandate that pollution must be eliminated. The EPA uses rather wide discretion in determining what degree of pollution control will be utilized on the basis of what is economically and technologically feasible. Officials admit that this makes the 1985 goal a focal point for planning and research, rather than an obligation to pollution-free water.

Effluent standards based on what control technology is able to achieve will permit introduction of low-level pollution. This is clear from data assembled by the EPA under the Industrial Wastes Studies Program. The material presents engineering judgments on achievable effluent values in 20 key industries according to the use of available control and treatment technology. For example, let us assume that available technology was used in integrated and petrochemical plants to refine all petroleum processed each year in the United States. According to the engineering considerations in the EPA study [14], 15 pounds of oil and 29 pounds of suspended solids per 1000 barrels of crude oil processed would still end up in effluent from the facilities. On the basis of 3,774,400,000 barrels of crude oil processed in U.S. refineries during 1968 [15], there would be a total of 28,308 tons of oil and 330,260 tons of suspended solids discharged into U.S. waterways each year, despite the use of available technology, if these petroleum plants processed the entire amount of crude oil. In the organic chemicals industry, based on estimated 1970 production, the manufacture of ethylene using available technology would still contribute 4500 tons of oil and 81,000 tons of dissolved solids to U.S. waterways every year. The manufacture of propylene would discharge wastes containing 360 tons of oil and 21,600 tons of total dissolved solids per year [16].

Less Visible Pollution

In addition, water pollution would continue even if pollution control technology required by the legislation achieved complete containment of all pollutants. This is because the legislation is aimed at identifiable pollution sources such as industrial facilities and municipal waste treatment plants—the predominant, but by no means the only, sources of water pollution. Agricultural activities, for example, account for about 20 percent of all stream pollution [17]. Drainage from construction, mining, and urban areas is a proportionately less significant source of pollution, since pollutants in these areas are not usually discharged through a pipe that can be conveniently monitored but instead seep into surface waters over wide areas and raise the level of contamination generally throughout watersheds.

The extent to which these sources of undetermined origin (so-called nonpoint sources) can predominate in some watersheds is illustrated by a study of three river systems in northern New Jersey—the upper Passaic, upper Raritan,

Floating garbage will be picked up by this debris control boat on the Chicago River, but the big problem is less visible pollution—frequently from undetermined sources.

and Millstone basins [18]. At the time of the study in 1970, some 750,000 people lived in the area, but urban and heavy industrial development were limited there. Members of the Water Resources Research Institute of Rutgers University wanted to devise a mathematical model to analyze the amount and origin of organic pollutants in the basins to aid in establishing water quality controls. The mathematical and computer manipulations necessary for the model proved less difficult than did the collection of accurate information about the origin of pollutants. On the basis of BOD levels, organic material in discharges from waste and industrial treatment plants in the Passaic River Basin accounted for an average of only 39 percent of the annual total organic load in the water over an 11-year period. As much as 37 percent came from undetected sources, which presumably included agricultural activities, septic tank discharges, drainage from city streets, spills, and illegal disposal of organic wastes. The balance of the organic content in the water was from natural contributions from soil, rocks, and vegetation. In the Millstone River Basin, authorized industrial and municipal discharges represented only 24.3 percent of the total organic load recorded in the water. The computed contribution from these authorized sources was lowest in the Raritan River Basin—only 7.3 percent of the total organic load. In light of their findings, the Rutgers scientists pointed out the futility of treating recorded effluents to achieve desired water quality without considering the effects of undesignated sources of pollution.

To counteract this kind of disparity, the 1972 federal law requires the development of area-wide waste management plans in regions that have severe nonpoint-source pollution problems. However, development and administration of the plans will be a state responsibility. The EPA will assist the states by preparing guidelines for control of pollution from agriculture, mining, construction, and underground waste disposal (the law does not deal extensively with deep-well disposal of industrial wastes, other than to require that states determine that such disposal is harmless before issuing permits for it).

An Alternative Approach

Even if both identifiable and elusive nonpoint-source pollution discharges were controlled under the new legislation, water pollution would still exist. We still would be left with many situations in which a product is manufactured cleanly in one place only to contaminate water many miles away during use or disposal. One example is the manufacture of ammonium nitrate fertilizer in Tennessee described previously in this chapter. Similarly, even wholly pollution-free petroleum refining or coal mining would create fuels that, during use, would produce air and water pollution in the form of gaseous or metallic by-products of combustion. The most careful manufacture of plastic disseminates products from which potentially toxic plasticizers vaporize to contaminate the air and water [19]. Scavaging of cadmium during metal plating and galvanizing keeps that toxic metal from plant waste water, but the plating itself releases cadmium to the air and hence to the water when scrap metal is melted down for reuse [20].

All this ensures that the 1985 goal of no pollution discharges into water will be elusive. This is not to say that the new legislation establishing the goal is not a landmark in our efforts to regulate the massive pollution produced by our society. It is that. The law attempts to deal with water pollution on a uniform basis across political boundaries by employing pollution control measures regarded as advanced both technologically and administratively. If applied forcefully, it will improve the quality of many waterways.

As our understanding of the seriousness of water pollution has increased over the years, our laws have become more comprehensive and more effective. This evolution is consistent with other historical trends in the United States during which, under political pressure, laws have been periodically rewritten to govern our activities in areas of intense public concern more tightly. Each new law has been built at least partially on the foundation of old laws, however, and, as pointed out repeatedly in this book, a number of crucial concepts and techniques in the old foundation are very weak.

An alternative approach would be to build a new foundation from our modern understanding of the importance of the hydrological cycle throughout the biosphere. Rather than focusing sharply on the most visible—albeit most serious—forms of direct discharge, recognition of the integrity of the hydrological cycle as paramount to life would necessitate adjustments in human activities to prevent contamination of that cycle at any point. (Some of the specific ways in which this could be done—such as recycling industrial and municipal wastewater—have been mentioned previously.) Also required would be ecologically sound means of utilizing the energy needed to power industry, transportation, and domestic utilities. Before a water pollution control structure could rise on its new foundations, much intensive research on new productive processes would have to be conducted. One reason for the lag in this research is that existing laws, including the new water pollution control legislation, emphasize refinement of current pollution control methods rather than ways to eliminate pollution by substituting a nonpolluting technology for a polluting one. To shift to a genuinely nonpolluting technology would require economic and social, as well as technological, changes in a system whose present goals, organization, and operation are not directed toward ecological integrity [21].

The extent to which future programs to eliminate water pollution go beyond the present painful step-by-step legislative struggle remains to be seen. One factor will be the intensity of public demand, based on an awareness of the bond between all life and the hydrological cycle. Another element will be the degree to which current water pollution control dogma is alerted to encompass biological insight into the environmental meaning of water pollution. Public and professional awareness has already progressed far since those early days in Chicago when the only sanitary goal was to pipe sewage far enough away to avoid human contact and then let dilution take care of the rest. We began with that era in the first chapter of this book; subsequent chapters have developed the scientific rationale for a new approach—one only partially realized in recycling projects and in the latest water pollution control law. However, given the history of the unregulated use of water for conflicting needs, these developments are encouraging signs that we are willing, even though grudgingly, to regulate our affairs to the imperative of nature.

Water pollution is a key indicator of our capacity to understand and adjust to that imperative. The pervasiveness of the hydrological cycle and the unique properties of water combine to capture large quantities of most pollutants, whether originally discharged to air, land, or water. In that sense, the purity of water is an indicator of the general quality of the environment. If we succeed in freeing water of pollution, we will have solved the environmental crisis itself. The purpose of this book has been to demonstrate this fact in the hope that one day the sensible reordering of our affairs may be mirrored in bright, pure water.

The earth's precious treasure of pure, fresh water is nowhere more beautiful than in her two deepest lakes—Lake Tahoe (above) in California and Lake Baikal (below) in Siberia. Tahoe has been threatened by sewage from the burgeoning recreational use of the lake; Baikal now has a pulp mill on its shores.

This complex plant now treats sewage from the Tahoe area and channels it down the mountain where it forms a new lake—pure enough for swimming. (1) Headworks and primary sedimentation basin (2) Primary sedimentation basin (3) Activated sludge aeration tanks (4) and (5) Secondary clarifier (6) Chemical flocculation basin (start of advanced treatment process) (7) Return sludge pump station (8) Chemical clarifier (9) Nitrogen removal tower (10) Separation beds and carbon columns (11) Backwash decant tank (12) Solids disposal building, housing sludge incinerator, lime-recalcining furnace, and dewatering facilities (13) Plant effluent pump station (14) Effluent storage ponds (15) Administration building (16) Sludge digesters (stand-by use only). Lake Baikal (below) still receives waste water from its pulp mill after treatment at a plant, one unit of which is shown here.

[1] William F. Jopling, David G. Deaner, and Henry J. Ongerth, "Fitness Needs for Wastewater-Reclamation Plants," *Journal of the American Water Works Association,* October 1971, Vol. 63, No. 10, 626–29.

[2] George A. Sawyer, "New Trends in Wastewater," *Chemical Engineering,* July 24, 1972, 120–28.

[3] K. D. Linstedt, K. J. Miller, and E. R. Bennett, "Metropolitan Successive Use of Available Water, and Henry J. Greaser, "Dallas Wastewater-Reclamation Studies," both *Journal of the American Water Works Association,* October 1971, Vol. 63, No. 10, 610–15 and 634–40, respectively.

[4] "AWWA Policy Statement on the Use of Reclaimed Wastewaters as a Public Water-Supply Source," *Journal of the American Water Works Association,* October 1971, Vol. 63, No. 10, 609.

[5] L. G. Suhr, "Some Notes on Reuse," *Journal of the American Water Works Association,* October 1971, Vol. 63, No. 10, 630–33.

[6] Ibid., p. 633.

[7] "Closing the Loop on Waste Waters," *Environmental Science and Technology,* August 1972, Vol. 6, No. 8, 692–93.

[8] *The Economics of Clean Water: Summary of Analysis,* Environmental Protection Agency (Washington, D.C.: U.S. Government Printing Office, 1972), pp. 1–33.

[9] "Industry Expenditures for Water Pollution Abatement," *The Economics of Clean Water, Volume 3,* Environmental Protection Agency (Washington, D.C.: U.S. Government Printing Office, 1972), p. 2.

[10] "Laws of the United States Relating to Water Pollution Control and Environmental Quality," compiled by the Committee on Public Works, U.S. House of Representatives (Washington, D.C.: U.S. Government Printing Office, July 1970).

[11] "Conference Report on S. 2770, Amending Federal Water Pollution Control Act," *Congressional Record* (Washington, D.C.: U.S. Government Printing Office, September 28, 1972), Vol. 118, No. 153, H8859–903.

[12] Ibid., p. H8860.

[13] Ibid., p. H8886.

[14] "Summary Report on the Petroleum Refining Industry," Industrial Wastes Studies Program, Environmental Protection Agency, Office of Water Programs, Division of Applied Technology, Washington, D.C., January 25, 1972.

[15] *Minerals Yearbook, 1968, Volume I–II,* U.S. Bureau of Mines (Washington D.C.: U.S. Government Printing Office, 1969), p. 821.

[16] Computations for ethylene and propylene are based on the following data: An estimated 1,800,000,000 pounds of ethylene and 7,200,000,000 pounds of propylene were produced in 1970 ("An Upbeat Year for Chemicals," *Chemical &*

Engineering News, September 6, 1971, 12a); with the use of available treatment technology, the waste discharge during production of these two chemicals contains .5 pound of free-floating oil and 9 pounds of dissolved solids per 1000 pounds of ethylene produced and .1 pound of free-floating oil and 6 pounds of dissolved solids per 1000 pounds of propylene produced *(Summary Report on the Organic Chemicals Industry,* The Industrial Wastes Studies Program, Environmental Protection Agency, Office of Water Programs, Division of Applied Technology, Washington, D.C., January 31, 1972).

[17] *The Cost of Clean Water, Volume II: Cost Effectiveness and Clean Water,* Environmental Protection Agency (Washington, D.C.: U.S. Government Printing Office, March 1971), p. 64.

[18] William Whipple, Jr., "BOD Mass Balance and Water Quality Standards," *Water Resources Research,* June 1970, Vol. 4, No. 3, 827–37.

[19] Kevin P. Shea, "The New-Car Smell," *Environment,* October 1971, Vol. 13, No. 8, 2–9.

[20] Julian McCaull, "Building a Shorter Life," *Environment,* September 1971, Vol. 13, No. 7, 2–14 and 38–41.

[21] The complex interrelationships between technology, pollution, and economic motivation are explored by Barry Commoner in *The Closing Circle* (New York: Alfred A. Knopf, 1971.)

APPENDIX

Table 1

Unless otherwise noted, results of interactions are derived from laboratory experiments on various species of aquatic life.

Results of Interactions with:

Substances	pH	Temperature	Hardness
Ammonia; Ammonium; Ammonium Hydroxide	Toxicity increases with increase in pH. In experiments with fish, toxicity increased 200% when pH was raised from 7.4 to 8.0.		
Cadmium			
Calcium			
Chlorine	Toxicity increases with decrease in pH.		
Chromium			
Cobalt Sulfate			
Copper	Above pH 7, a precipitate—copper hydroxide—may be formed, making the solution less toxic.		Copper precipitates are formed in hard water, reducing their toxicity.
Cyanide; Hydrogen Cyanide	Below pH 8, hydrogen cyanide is in its undissociated form (HCN) and is toxic. When the pH of a solution containing a nickelo-cyanide complex dropped from 8.0 to 6.5, the toxicity of the solution increased 1000 times [7].	Toxicity is increased two to three times with a 10°C rise in temperature.	In some experiments, CN was more toxic in soft water.

Results of Interactions with: Oxygen	Other Chemicals	Possible Sources of Pollutant
Low oxygen increases toxicity.	Carbon dioxide reduces toxicity by lowering the pH.	Chemical, gas, and ice plants; scouring and cleaning operations using ammonia water.
	Synergistic with copper. The toxicity of zinc and cadmium is additive [1].	Electroplating; ceramics pigmentation; insecticides; textile printing.
	Antagonistic to lead, zinc, aluminum, and toxic solutions of sodium, magnesium, and potassium chloride salts [2].	Brewing; tanning; production of mineral water; manufacturing of glue, paint, and explosives; runoff from roads.
	Toxic substances are formed by the addition of chlorine to a variety of chemicals: combines with thiocyanate to transform a nontoxic effluent sample into a toxic one [3]; converts parathion to paraoxan, a more toxic insecticide [4]; combines with the phenols and ammonia to form chlorophenols and chloramines, both toxic substances [5].	Sewage treatment; bleaching (as in textile mills); chemical control of biological growth in cooling waters.
Toxicity of sodium chromate increases as dissolved oxygen decreases.		Metal pickling; anodizing aluminum; tanning; manufacture of paints, dyes, explosives, ceramics.
	Increases the toxicity of nickel.	Decorating and plating.
Toxicity increases with decreases in oxygen tension.	Synergistic with chlorine, zinc, cadmium, and mercury. Exhibits antagonism toward sodium nitrite and sodium nitrate, and decreases the toxicity of cyanide.	Corrosion of copper tubing; algicides; textile pigmentation; tanning; photographic processing; engraving; electroplating.
Toxicity increases with decreases in oxygen tension [6].	Becomes very toxic when mixed with zinc or cadmium.	Gas works; coke ovens; steel (gas scrubbing); metal cleaning and electroplating; chemical industries.

Results of Interactions with:

Substances	pH	Temperature	Hardness
Hydrochloric Acid	Toxic only when pH is below 5.		
Maganese			
Mercuric Chloride			
Nitriloacetic Acid (NTA)			
Organochlorine Insecticides			
Phenol		Toxicity increases with increasing temperature.	Toxicity increases with decreasing hardness.
Propanil (an aniline-based herbicide)			
Sodium			
Sodium Ferrocyanide			

Results of Interactions with:		Possible Sources
Oxygen	Other Chemicals	of Pollutant
	Nickel and manganese are antagonistic.	Steel alloys; dry-cell batteries; glass; ceramics; paints and varnishes; inks; dyes; fireworks; agriculture.
	Synergism results when small amounts of sodium chloride are added to mercuric chloride; the effect becomes antagonistic if high concentrations of sodium chloride are added.	Embalming; disinfecting; preserving; fabric printing; tanning; electroplating; manufacture of ink.
	Synergistic with cadmium. In experiments with rats, teratogenicity increased ten times when NTA and cadmium were combined [8].	Used temporarily in detergents as a substitute for phosphate builders.
	In experiments with mammals, it has been found that various insecticides interact with each other in biological tissue. For example, when aldrin, dieldrin, endrin, heptachlor, or chlordane are fed to animals in conjunction with DDT, the amount of these insecticides that is stored (mainly in the liver) is greatly reduced. A number of drugs, such as the barbituates, are also more quickly metabolized (broken down and passed through the body) when administered with DDT [9].	
Toxicity increases with decreasing oxygen.		Disinfectants; manufacture of synthetic resins; medical and industrial chemicals; distillation of wood; gas works; coke ovens; oil refineries; chemical plants.
	One of the degradation products of propanil in the soil is TCAB (3,3′, 4,4′–tetrachlorazobenzene); bacterial carry out the degradation. TCAB is not carcinogenic, but the same type of microbiological breakdown could similarly transform related herbicides to azo-compounds which have been shown to be carcinogenic [10].	
	Antagonistic toward the toxicities of potassium, ammonium, and calcium salts.	
	Decomposes under sunlight, producing cyanide and hydrogen cyanide. Under these conditions, sodium ferrocyanide is more toxic.	Blueprinting; photographic processing; pigmentation and dyes.

Results of interactions with: Substances	pH	Temperature	Hardness
Sodium Sulfide	As pH decreases from 9 to 6, toxicity increases.		
Sulfuric Acid	Toxic when pH is below 5.		
O-Xylene		Toxicity triples with an increase in temperature of 8°C.	
Zinc		Toxicity increases with increasing temperature.	Toxicity decreases with increasing hardness.

[1] Peter Dudoroff and Max Katz, "Industrial Wastes: Critical Review of Literature on the Toxicity of Industrial Wastes and Their Components to Fish. II. The Metals, as Salts," *Journal of the Water Pollution Control Federation,* July 1953, Vol. 25, No. 7, 811.

[2] Ibid., p. 832.

[3] H. A. Hawkes, "Biological Aspects of River Pollution," Chapter 8 in *River Pollution: Causes and Effects,* Volume II (London: Butterworth, 1962), p. 337.

[4] S. K. Kushnaswami, "Health Aspects of Water Quality," *American Journal of Public Health,* November 1971, Vol. 61, No. 11, 2262.

[5] John W. Arthur and John G. Eaton, "Chloramine Toxicity to the Amphipod *Gammerus pseudolimnaeus* and the Fathead Minnow *(Pimephales promelas),*" *Journal of the Fisheries Research Board of Canada,* 1971, Vol. 28, No. 12, 1841–45. *See also* Terri Aaronson, "Spectrum," *Environment,* January-February 1972, Vol. 14, No. 1, 25.

[6] *Water Quality Criteria,* report of the National Technical Advisory Committee to the Secretary of the Interior, Federal Water Pollution Control Administration, Washington, D.C., April 1, 1968, p. 65.

[7] Ibid., p. 63.

Results of Interactions with: Oxygen	Other Chemicals	Possible Sources of Pollutant
		Tanning; beet-sugar processing.
		Production of fertilizers, acids, dyes, explosives.
		A solvent and cleaning agent; used in the manufacture of dyes and organics.
Toxicity increases as dissolved oxygen is lowered.	Synergistic with nickel, cyanide, and copper. The toxicity of zinc and cadmium is additive.	Galvanizing; in alloys; printing plates; dye-manufacture; paint pigmentation; cosmetics; pharmaceuticals; insecticides.

Source: Most information in this table is from Jack E. McKee and Harold W. Wolf, *Water Quality Criteria*, State Water Quality Control Board, Sacramento, California, 1963. Other reference sources are noted.

[8] Barry Commoner, *The Closing Circle* (New York: Alfred A. Knopf, 1971), p. 224.

[9] Joseph Street, "Organochlorine Insecticides and the Stimulation of Liver Microsome Enzymes," in *Biological Effects of Pesticides in Mammalian Systems, Annals of the New York Academy of Sciences,* June 23, 1969, Vol. 160, Art. 1, 274–90.

[10] David Pramer, "The Soil Transforms," *Environment,* May 1971, Vol. 13, No. 4, 42–46.

Table 2

Contaminant	Recommended Limit (in mg/l unless otherwise specified)	Number of Communities Exceeding Recommended Limit	Mandatory Limit (in mg/l)	Number of Communities Exceeding Mandatory Limit
Biological Coliform			*	120
Organic Chemical				
CCE (carbon-chloroform extract)[a]	0.2	12		
Inorganic Chemical				
Arsenic	0.01	2	0.05	2
Barium			1.0	1
Boron[b]	1.0	9	5.0	
Cadmium			0.01	3
Chloride	250.0	9		
Chromium (hexavalent)			0.05	4
Copper	1.0	11		
Fluoride	0.8–1.7[c]	52	1.4–2.4	24
Iron	0.3	96		
Lead			0.05	14
Manganese	0.05	90		
Nitrate	45.0	19		
Selenium			0.01	5
Sulfate	250.0	25		
Zinc	5.0	1		
Particulate				
Total dissolved solids	500	95		
Turbidity in untreated water	5 units	26		
Turbidity in treated water	1 unit	26		
Radioactive				
Radium 226[d]	3 pC/l	6		
Color	15 units	7		

*Arithmetic average coliform density no greater than one per hundred milliliters, or two or more samples (five percent if more than 20 examined) greater than four per hundred milliliters when using membrane filter technique.

[a] Water is first filtered through activated charcoal and organics are extracted in chloroform. Only 110 of the 969 water systems in the study were tested for organic chemicals.

[b] Not included in 1962 PHS standards, but proposed.

[c] Standard varies with temperature.

[d] Radium 226 is a naturally occurring radioisotope found in the water in areas where the rock is high in radium. Radioactivity is measured in curies, or fractions thereof. A picocurie (pC) is a millionth of a millionth of a curie.

Photo Credits and Copyright Acknowledgments

CHAPTER 1
Pp. 2–3, "Lake Michigan Beach," photograph by Brett Weston, Rapho-Guillumette.
P. 4, photograph by John Hendry, Jr., National Audubon Society.
P. 5, photograph courtesy of the Michigan Department of Natural Resources.
P. 10, photograph by Dave Healey.
P. 22, graph by W. F. Carbine, reprinted from *Conference on the Matter of Pollution of Lake Michigan and Its Tributary Basins*, Sessions 1–3, Environmental Protection Agency (Washington, D.C.: U.S. Government Printing Office), p. 1072. Photograph by Geoffrey Gove, Rapho-Guillumette.

CHAPTER 2
Pp. 30–31, photograph by Brett Weston, Rapho-Guillumette.
P. 34 (top), photograph by Gene Ahrens.
P. 34 (bottom left), *Environment* photograph by Robert Charles Smith.
P. 34 (bottom right), *Environment* photograph by Robert Charles Smith.
P. 35 (bottom left), *Environment* photograph by Robert Charles Smith.
P. 35 (right), photograph by Ray Atkeson.
P. 39, graph by John Cairns, Jr., reprinted from *Environment*, October 1968, p. 191. Photographs (left, center) © Biological Supply House, Inc., Chicago, Illinois; photograph (right), Harbrace photo.
P. 40, graph by John Cairns, Jr., reprinted from *Environment*, October 1968, p. 191.
P. 43, photograph by Harvey Stein.
P. 44, photograph by Bucky and Avis Reeves, National Audubon Society.
P. 52, *Environment* photograph by Daniel T. Magidson.
P. 55, graph by Gene E. Likens, F. Herbert Borman, and Noye M. Johnson, reprinted from *Environment*, March 1972, p. 39.
P. 56, photograph courtesy of the Missouri Department of Conservation.
P. 62, photograph courtesy of the Japan National Tourist Association.
P. 63, photograph courtesy of the Japan National Tourist Association.

CHAPTER 3
Pp. 68–69, photograph by Chuck Ashley, Rapho-Guillumette.
P. 71, cable photograph courtesy of UPI.
P. 72, photograph courtesy of Exxon.
P. 76, photograph courtesy of the Office of Public Affairs, Environmental Protection Agency.
P. 80, photograph courtesy of the U.S. Signal Corps.

CHAPTER 4
Pp. 88–89, photograph courtesy of the Freeport Sulphur Company.
P. 91, *Environment* photograph by Dave Healey.
P. 94 (top), photograph courtesy of the American Iron and Steel Institute, Washington, D.C.
P. 94 (bottom), photograph courtesy of the Bethlehem Steel Corporation.
P. 97, Harbrace photo.
P. 105, photograph by J. Allan Cash, Rapho-Guillumette.
P. 107, photograph courtesy of the U.S. Department of Agriculture.
P. 110, photograph by Harvey Stein.
P. 112, graph from *The Cost of Clean Water and Its Economic Impact*, Volume I (Washington, D.C.: U.S. Government Printing Office, 1969), p. 44. Photograph by J. Pavlovsky, Rapho-Guillumette.
P. 117, *Environment* photograph by Robert Charles Smith.

CHAPTER 5

Pp. 122–123, "Water Pollution in Lake Michigan: Chicago's Belmont Harbor, July 1968," photograph by John Hendry, Jr., National Audubon Society.
P. 124, photograph by Ray Montgomery.
P. 125, photograph courtesy of Dudley, Hardin, and Yang, Inc.
P. 127, graph adapted from *Conference on the Matter of Pollution of Lake Michigan and Its Tributary Basins,* Sessions 1–3, Environmental Protection Agency (Washington, D.C.: U.S. Government Printing Office).
P. 131, photograph by Harvey Stein.
P. 133, graph by Stig H. Fonselius, reprinted from *Environment,* July–August 1970, p. 7.
P. 135, photograph by Don Wooldridge, courtesy of the Missouri Department of Conservation.
P. 140, *Environment* photograph by Dave Healey.

CHAPTER 6

Pp. 148–149, photograph by Robin Forbes.
P. 157 (top), courtesy of *Environment.*
P. 157 (bottom), photograph by Bruce Lund, Massachusetts Audubon Society.
P. 159, photograph by Dave and Lyn Hancock, Island View Wildlife Research Center, Saanichton, British Columbia, Canada.
P. 163, both photographs courtesy of the Institute of Environmental Education.
P. 168, photograph courtesy of the Lake Michigan Federation.

CHAPTER 7

Pp. 174–175, photograph by Tim Eagan.
P. 177, photograph courtesy of the Farmers Chemical Association.
P. 181, photograph by Robin Forbes.
P. 183, photograph courtesy of the U.S. Army Corps of Engineers, New York District.
P. 186 (top), photograph courtesy of CH2M/Hill.
P. 186 (bottom), photograph by B. Klinchenko, APN, Novosti Press Agency.
P. 187 (Top), photograph courtesy of CH2M/Hill.
P. 187 (bottom), photograph courtesy of the Novosti Press Agency.

P. 197, The bottom of McMurdo Sound, Antarctica, near U. S. McMurdo Station. Depth, 100 feet. Photograph by Bill Curtsinger, Rapho-Guillumette.

Italic numbers refer to illustrations and photographs.

Index

A 3
B 4
C 5
D 6
E 7
F 8
G 9
H 0
I 1
J